THE GEOLOGY OF THE SOUTHERN VANCOUVER ISLAND

A Field Guide

C.J. YORATH & H.W. NASMITH

(With contributions by A. Sutherland Brown, N.W.D. Massey and D.F. VanDine)

ORCA BOOK PUBLISHERS

First printing, 1995

Canadian Cataloguing in Publication Data
Yorath, C.J., 1936-
The geology of Southern Vancouver Island

Includes bibliographical references and index.
ISBN 1-55143-032-0

1. Geology, Structural–British Columbia–Vancouver Island.
2. Geology–British Columbia–Vancouver Island. I. Nasmith, Hugh. II. Title.
QE187.Y67 1995 557.11'2 C95-910208-6

Financial assistance provided by the Canadian Geological Foundation
Cover design by Christine Toller
Cover photograph by Russ Heinl
Printed and bound in Canada

Orca Book Publishers
PO Box 5626, Station B
Victoria, BC Canada
V8R 6S4

TABLE OF CONTENTS

FOREWORD

Here is a field guide that breaks new ground. You hold a book focused on the geology of Greater Victoria, also with an introductory story on the origins of Vancouver Island. Good bookstores offer rows of field guides on subjects ranging from wildflowers to birds and from stars to butterflies, many of them bestsellers, but you won't find another guide like this among them.

This one leads you into local places to show you rocks made hundreds of million of years ago, and into valleys made only yesterday (geologically speaking) when glaciers were grinding down Vancouver Island's rocky surface. Somehow the land takes on new meaning when you know why Mt. Douglas rises above the surrounding flats and how more recent events dug out the long inlet that creates the Saanich Peninsula. Have you learned to see the evidence of glaciers on rocks throughout our area? Even the rock outcrop in my basement has a groove across it, ground out by rocky debris in the base of a moving glacier of immense weight.

Don't be afraid of geology. It is one of the great detective stories of our time, and this book shows you where to see the clues that reveal the very origins of the ground you stand on. With a little understanding of the shapes of landscapes and the events that shaped them, you will soon have new skills with which to read the information in landscapes both at home and on distant travels.

The authors have done an outstanding job of making geology as easy as words can make it. Browse through this book and you will find yourself planning family expeditions, or hikes with friends, so you can see — for yourself — into the past of this remarkable part of this remarkable planet.

Where to go, how to get there, and what to see there are all laid out in this book.

— Yorke Edwards

PREFACE

The preparation and publication of this guidebook was sponsored by the Pacific Section of the Geological Association of Canada. It is intended as a field guide for teachers, interested travellers, natural history buffs and geologists unfamiliar with this region. The book provides information on the geological history and architecture of the bedrock underlying the city of Victoria and surrounding areas, as well as on the history and products of glaciation which affected this region.

The guidebook is divided into two sections. Part One describes the landscape of Vancouver Island, its structure and geologic history, the effects of glaciation, the history of mining on the island, and the causes, nature and significance of earthquakes which occur frequently throughout the region. Part Two describes the geology of many readily accessible localities throughout Greater Victoria and adjacent areas, particularly those at public parks and beaches. Locality descriptions are commonly cross referenced to relevant sections in the first part of the book, thus providing a broader context within which these descriptions can be understood.

Throughout the text technical language is used in explanatory context. Where such terms are first introduced they appear in **bold**-face type, indicating that their definitions are included in a glossary following the text. Formal names of geological features (formations, faults, terranes, etc.) appear in *italics* to aid the reader in scanning the text. On the inside of the front cover we provide the geological time scale as well as a column illustrating the geological succession and ages of the rock formations that occur on southern Vancouver Island.

To further enhance the usefulness of this guide the authors recommend the purchase of a street map of the Greater Victoria area. Also available, either through the map sales office of the British Columbia Government, Crown Publication Inc., or the publications office of the Geological Survey of Canada in Vancouver (see *Additional Sources*), are the following National Topographic Series maps for the Greater Victoria area:

Scale 1:50,000
1. Sooke. NTS 92B/5
2. Victoria. NTS 92B/6
3. Sidney. NTS 92B/11
4. Shawnigan Lake. NTS 92B/12
5. River Jordan. NTS 92C/8
6. Port Renfrew. NTS 92C/9

Scale 1:250,000
1. Victoria. NTS 92B
2. Cape Flattery. NTS 92C

Also, for the real keeners, we recommend purchase of the geological map of the Victoria area by J.E. Muller (GSC Map 1553A; Scale 1:100,000; available from the Geological Survey of Canada in Vancouver (see *Additional Sources*).

ACKNOWLEDGEMENTS

During the past century many individuals have contributed information used in the content of this guidebook. Among them are: N. Alley, A. Andrews, R.L. Armstrong, A. Blais, P.T. Bobrowski, M.T. Brandon, B.E.B. Cameron, D.J.T. Carson, C.H. Clapp, R.M. Clowes, J.J. Clague, D.S. Cowan, C.H. Crickmay, A. Desrochers, T.J.E. England, L.H. Fairchild, J.T. Fyles, H.C. Gunning, D. Howes, T. Hamilton, R.D. Hyndman, E. Irving, C. Isachsen, W.G. Jeffery, J.A. Jeletzky, S. Juras, N.W.D. Massey, A. McGugan, W.G. Milne, J.E. Muller, M.T. Orchard, J.A. Pacht, R.R. Parrish, R.P. Riddihough, G.C. Rogers, M.E. Rusmore, D.N. Shouldice, J.S. Stevenson, R.C. Surdam, A. Sutherland Brown, E.T. Tozer, J.L. Usher, D.F. VanDine, P.D. Ward, R.W. Yole and C.J. Yorath. Thanks are extended to the **Geological Association of Canada** for funding the drafting of diagrams and to the **Geological Survey of Canada** for many support services. The line drawings were drafted by E.G. Yorath and R. Franklin. Several people read various stages of the manuscript: technical reviews were provided by J.T. Fyles and C. Lowe; for readability we are grateful to L.C. Yorath, L.K. Law, R.D. McLean and A.H. Walters. C.E. Kilby, representing the Pacific Section of the Geological Association of Canada, conducted the final edit of the manuscript prior to its submission to Orca Book Publishers for publication.

The Pacific Section of the Geological Association of Canada thanks the Canadian Geological Foundation for its grant to defray the costs of publication of this guidebook. It also expresses gratitude to members of the planning committee of the Geological Association of Canada's annual meeting in Victoria (1995) for their assistance.

INTRODUCTION

Vancouver Island is said to belong to the **Insular Belt**, the westernmost of five northwesterly-trending subdivisions of the **Canadian Cordillera** (see diagrams on pages 6 and 18). The *Canadian Cordillera* consists of all of the systems of mountain ranges and plateaux extending from the International Boundary to the Arctic Ocean and from the foothills of Alberta to the toe of the **continental slope** off the west coasts of Vancouver Island and the Queen Charlotte Islands. The five belts, distinguished from one another through differences in rock type, internal structure and physiography, are a reflection of the architectural history of the *Cordillera*. The *Insular Belt*, consisting of Vancouver Island, the Queen Charlotte Islands and parts of southeastern Alaska and southwestern Yukon, is a product of collisions between the ancient western edge of North America and exotic pieces of **crust**, including Vancouver Island, which formed in far away places and which collided with the continent about one hundred million years ago. The piece of crust that included Vancouver Island is known as **Wrangellia.**

The geological history of Vancouver Island is one of development by volcanic eruptions. Throughout some 375 million years, three major periods of volcanism, separated by episodes of sediment accumulation, built an edifice which, after it joined with the westward-moving North American continent, was subjected to forces that compressed and fractured the rocks when other exotic fragments of the earth's crust were rammed against and beneath it. Later, when the earth's climate cooled, beginning about three million years ago, thick glacial domes coalesced to form a mass of ice that covered all of western Canada's mountain systems, including those on Vancouver Island. **Fiords**, river valleys, the shapes of mountains, scratches and grooves on rocks, and the thick piles of sand and gravel along our coasts all bear testament to a time when Vancouver Island lay encased in ice. Through the processes of plate tectonics and sea-floor spreading, volcanism, sediment accumulation, mountain building, glaciation and erosion, Vancouver Island has come to its present form.

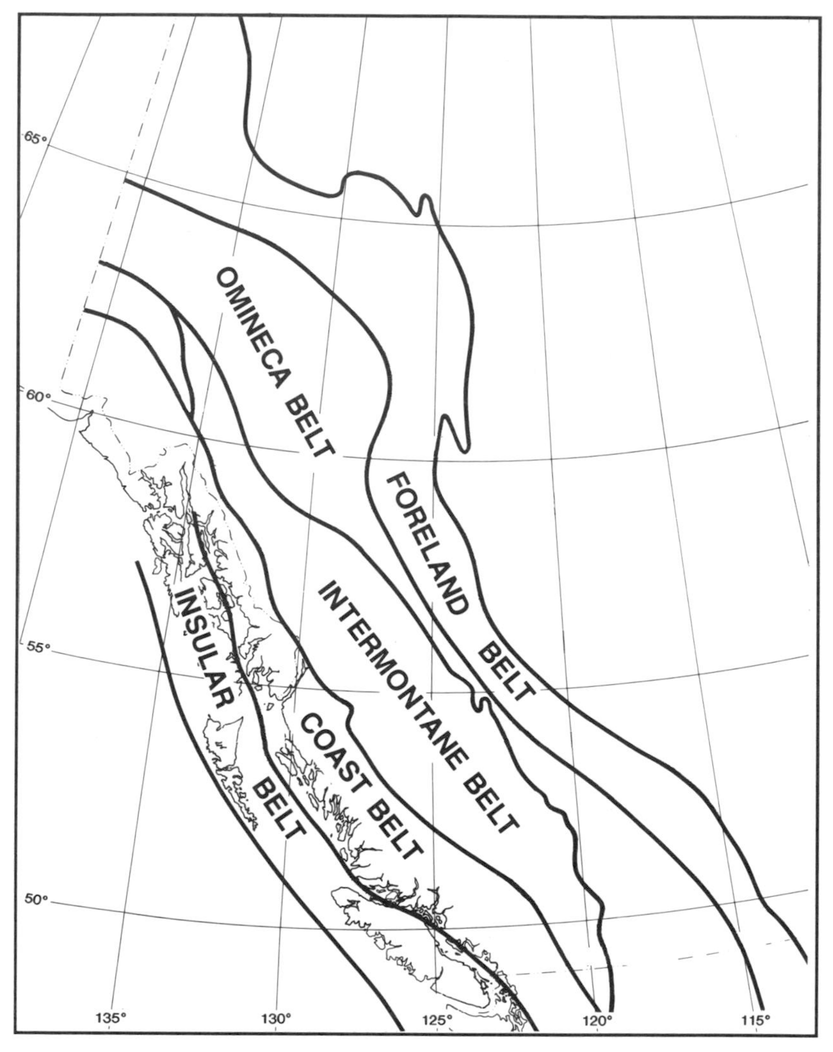

The Canadian Cordillera is shown here to be subdivided into five zones, or belts, which extend throughout its length. On the east is the Foreland Belt, consisting almost entirely of sedimentary strata exposed in the Rockies, Mackenzies, Franklins, Ogilvies and Richardson Mountains. The Omineca Belt, composed of sedimentary, igneous and metamorphic rocks, contains the rugged Purcells, Selkirks and Cariboos, as well as the Omineca, Cassiar and Selwyn Mountains. The Intermontane Belt is mainly a region of subdued, rolling and plateau topography underlain dominantly by volcanic and sedimentary rocks. The Coast Belt has been moulded from granitic and metamorphic rocks and contains some of the most spectacularly rugged mountains on the continent. The Insular Belt is composed of granitic, volcanic, sedimentary and metamorphic rocks, which form low but rugged mountains. In the north the Insular Belt contains the spectacular Saint Elias Mountains, which have the highest and grandest peaks in Canada.

PART ONE

The Geological Architecture and Origin of Vancouver Island

THE LANDSCAPE

If we were in a space vehicle orbiting the earth, we would be able to see all of Vancouver Island at a glance. We would see that it is located about midway along and just off the west coast of North America, and that it is the largest island anywhere along the coast. The island is about 450 kilometres (km) long and is oriented in a northwest-southeast direction. It lies between 48°20' N and 50°40' N, and 123°10' W and 128°30' W. Although 125 km at its widest point, the average breadth of the island is only 70 km. The total land area is approximately 32,000 square kilometres.

From the space vehicle we also would see that the island has a very irregular shoreline, particularly along its west coast. There are many long inlets and innumerable smaller islands surrounding the main island. Alberni Inlet is the longest inlet, extending 60 km inland from the west coast and leaving Vancouver Island only 25 km wide at that point. To circumnavigate the island we would have to travel at least 600 nautical miles. However, if we wanted to closely follow the irregular coastline, we would have to travel twenty to twenty-five times that distance, essentially back and forth across Canada, from Victoria to St. John's, Newfoundland, twice!!

If we were in an airplane, flying closer to the surface, we would get a more detailed view of the island, including its third dimension, or **relief**. The relief seen would vary from relatively low flatlands along the east and west coasts to rugged mountains, rising to 2,200 metres (m) above sea level, in the central interior. Vancouver Island can be divided into several **physiographic regions**, based upon the amount of relief, topographic complexity and landscape characteristics.

The physiographic regions of Vancouver Island fall into three broad categories: mountains, highlands and plateaux, and lowlands and basins.

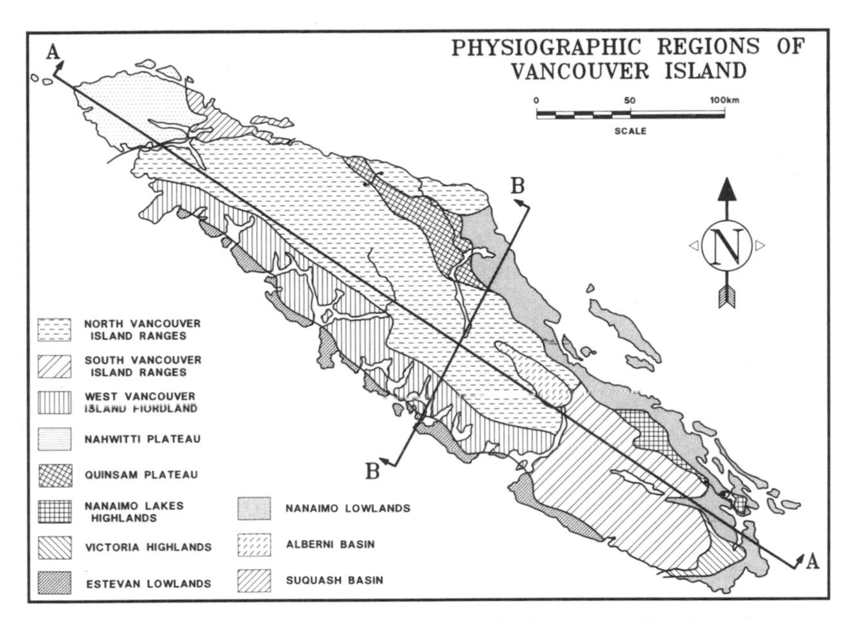

Physiographic subdivisions of Vancouver Island. The two profiles show the variation in elevation along the lines A-A and B-B.

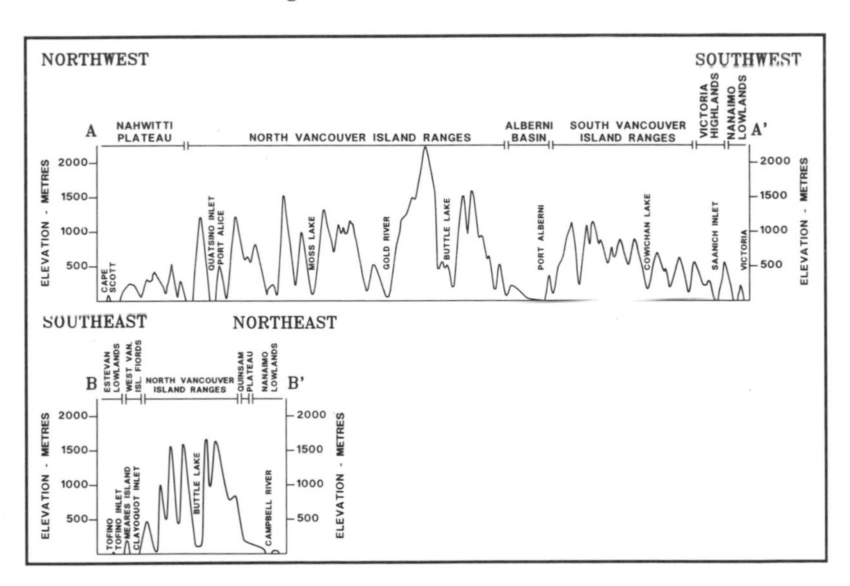

MOUNTAIN REGIONS

North Vancouver Island Ranges

The *North Vancouver Island Ranges* physiographic region is the largest and most scenically spectacular on Vancouver Island. It extends as a sixty-kilometre-wide area of rugged mountains, 270 km in length, from Quatsino Sound and the Suquash Basin in the north to Alberni Inlet and Port Alberni in the south and also includes the rugged coastal region north of the Quinsam Plateau. Forming the core of the island, the region consists of numerous northwest-southeast trending mountain ranges dissected by deep valleys and containing the highest mountains on the island. At 2,200 m in elevation, The Golden Hinde is the highest, followed by Elkhorn Mountain (2,192 m), Mt. Victoria (2,164 m) and Mt. Colonel Foster (2,135 m). Many peaks are snow-capped well into the summer, and a few have permanent glaciers. Summit elevations are greatest in the south-central portion of the region in the vicinity of Strathcona Park and diminish towards the north and west. The *North Vancouver Island Ranges* also boast Della Falls (see photo on page 11) in the southern portion of Strathcona Park. At 440 m, Della Falls is the highest waterfall in Canada.

South Vancouver Island Ranges

The *South Vancouver Island Ranges* physiographic region is smaller in area and less visually dramatic than its northern counterpart. These lower and less rugged mountains, covering an area approximately 100 km by 70 km, lie southeast of Alberni Inlet. Cowichan Lake roughly divides the region in two. Northeast of the lake the mountain peaks rise 1,500 m to 1,800 m, while to the southeast the highest summits range from 900 m to 1,400 m. Some of the better-known peaks in this region include Mt. Arrowsmith (1,818 m) (see photo on page 12), Mt. Moriarty (1,611 m), Mt. Whymper (1,542 m) and Green Mountain (1,466 m). Glaciers are not found in these mountains; however, climbers should be prepared for snow at any time of year at upper elevations.

West Vancouver Island Fiordland

Moderately high mountains, dissected by numerous long and narrow fiords, characterize the *West Vancouver Island Fiordland*. This physiographic region extends for 260 km along the western side of the island, from the Brooks Peninsula to Alberni Inlet. The inlets are bordered by steep-sided, rocky slopes which continue well below

The cataract of Della Falls in western Strathcona Provincial Park cascades some 440 m over granitic rocks of the North Vancouver Island Ranges.

sea level to form deep saltwater channels. They generally cut through the mountains at an angle approximately perpendicular to the coastline and extend 10 to 40 km inland. The fiords are the remains of old river valleys which were deepened by glaciation and flooded by the sea. Besides Alberni Inlet, which reaches inland for 60 km and is 330 m deep at its deepest point, other well-known inlets include Tofino Inlet, Chuchalat Inlet and Tahsis Inlet.

HIGHLANDS AND PLATEAUX

Nahwitti Plateau

Nahwitti Plateau occupies the northwestern tip of Vancouver Island. The eastern two-thirds of this physiographic region is characterized by a gently rolling terrain with rounded hills, generally less than 600 m in elevation. Further to the west, broad lowlands and valleys commonly characterize the rolling plateau. You cross this plateau and several broad lowlands as you drive from Port Hardy to Holberg then on towards Cape Scott Provincial Park.

This northerly view across the South Vancouver Island Ranges (foreground) and adjacent Nanaimo Lakes Highland (middle background) focuses on Cameron Lake and on the lava pile of Mt. Arrowsmith.

Quinsam Plateau

Quinsam Plateau is a region of moderate relief, nestled between two areas of the North Vanc*ouver Island Ranges*. It extends in a northwest-southeast direction for approximately 100 km from Johnstone Strait in the north almost as far south as Courtenay. Moderate to rugged hills, ranging in elevation between 200 and 1,000 m, are characteristic of the region. The southern portion is dotted by numerous lakes, including Campbell Lake and the Quinsam series of lakes. The northern portion is drained principally by the Adam River.

Nanaimo Lakes Highland

Nanaimo Lakes Highland is a region of transition between the *South Vancouver Island Ranges* and the *Nanaimo Lowland*. This physiographic region extends for 50 km in a northwest-southeast direction west of the city of Nanaimo. A small isolated area of this region occurs on Saltspring Island, where it is represented by Mt. Maxwell. Most of the landscape varies between 200 and 1,000 m in elevation, although towards the west a few higher mountainous areas reach beyond 1,000 m above sea level. Most of the region is drained by the Nanaimo Lakes and the Nanaimo River and its tributaries.

From a viewpoint on the Malahat, the glacial fiord of Finlayson Arm extends southward between the rounded hills of the Victoria Highland.

Victoria Highland

Similar to the *Nanaimo Lakes Highland*, *Victoria Highland* physiographic region is a geographic transition between the *South Vancouver Island Ranges* and the *Nanaimo Lowland*. The *Victoria Highland* extends from just south of the city of Duncan, wrapping around the South Vancouver Island mountains as far west as River Jordan. This region includes the Malahat area along the west side of Saanich Inlet and the Highlands district of the Greater Victoria area. Most of this physiographic region is between 200 and 500 m in elevation, with a few low mountains greater than 500 m. Lakes are common in this region, the most familiar being Shawnigan Lake and Sooke Lake. Most of the Greater Victoria Water District is located within the *Victoria Highland*.

LOWLANDS AND BASINS

Estevan Lowland

Estevan Lowland occupies a narrow region stretching approximately 290 km along the southwest coast of Vancouver Island. It is dissected by numerous inlets and fiords and extends from just south of Brooks Peninsula to Port Renfrew. The region is commonly less than 3 km wide, except on Hesquiat Peninsula, where it is 12 km. Generally, the *Estevan Lowland* is flat and featureless, elevations

An excellent view of the Nanaimo Lowland can be obtained from the top of Mt. Prevost near Duncan.

being commonly less than 50 m. In places the subdued topography is interrupted by small, irregular, steep-sided hills that may reach 100 m above sea level. Most of the coastline of the *Estevan Lowland* is very rocky. Except for the abundance of beaches, the terrain between Ucluelet and Tofino is typical of this physiographic region.

Nanaimo Lowland

The *Nanaimo Lowland* on the east side of the island extends from Campbell River in the north to Victoria in the south and includes most of the Gulf Islands. The width of the region is variable, from less than 2 km in the vicinity of Fanny Bay to approximately 35 km between Galiano Island and the upper Cowichan River valley. This physiographic region consists of gently rolling hills that give way to flatter plains bordering much of the Strait of Georgia. The hills reach elevations of 200 m above sea level. Approximately 90 percent of the population of Vancouver Island and most of the island's farmland and major transportation routes are contained within the *Nanaimo Lowland.*

Alberni Basin

Alberni Basin is the smallest physiographic region on Vancouver Island. It is a spoon-shaped lowland surrounded on all sides by mountains. The long axis of the spoon stretches for 40 km northwest of Port Alberni. The width varies from 8 to 13 km. This region, which does not exceed 200 m in elevation, is drained by the meandering Stamp and Somass rivers and contains some of the best farmland on the island.

Suquash Basin

Suquash Basin is a small, triangular-shaped lowland near the north end of Vancouver Island and includes Malcolm Island and several smaller islands in Queen Charlotte Strait. The region is characterized by gently rolling and level terrain, mostly under 300 m in elevation, although a few higher, rounded hillocks are scattered throughout the basin. The landscape between Port Hardy and Port McNeill is typical.

THE ORIGIN OF VANCOUVER ISLAND

INTRODUCTION

The origin and geological evolution of Vancouver Island is closely linked to that of the Queen Charlotte Islands and parts of southeastern Alaska, which together comprise a piece of the earth's crust known as **Wrangellia**. The concept of *Wrangellia* forms part of a broader hypothesis which attempts to explain how much of British Columbia and the Yukon originated as a series of separate, exotic **terranes**, or pieces of crust, that originated in distant latitudes in an ancient Pacific Ocean, and which were accreted to the edge of North America throughout the past 170 million years. Each terrane has its own unique geological history which is different from that of neighbouring terranes. *Wrangellia* and the **Alexander Terrane** together make up what we call the **Insular Superterrane**, a large piece of the earth's crust which, in addition to Vancouver Island and the Queen Charlotte Islands, includes the Alexander Archipelago of southeast Alaska and several isolated localities throughout the Coast Mountains and St. Elias Mountains, the latter situated in northwestern British Columbia and southwestern Yukon. Except for small areas in the south and west, most of Vancouver Island belongs to *Wrangellia*.

It should be understood that Vancouver Island has not always existed in the shape and form it has today, nor has it always been an island. Throughout the 375 million years of its history it has existed in many forms: as part of a broad submarine plateau; as chains of erupting volcanoes; and as several basins in which thick successions of sedimentary strata accumulated. Until about 100 million years ago, the "island," existing as a part of *Wrangellia*, was separate from North America, to which it became attached during the Cretaceous Period of the **Mesozoic Era**. Throughout the succeeding 100 million years, episodes of mountain building, erosion, glaciation and the accretion of additional terranes have moulded this part of *Wrangellia* into the Vancouver Island that we know today.

The following discussion of the origin of Vancouver Island is brief; some details are omitted. For additional information, refer to *Additional Sources* near the end of this guidebook.

THE PALEOZOIC ERA

The Devonian Period

The geological construction of Vancouver Island began during the **Devonian Period** of the **Paleozoic Era**, some 380 million years ago. Throughout much of its history Vancouver Island, together with parts of southeastern Alaska, the Queen Charlotte Islands and the southern Coast Mountains, all of which are parts of *Wrangellia*, moved as part of one or more crustal plates that make up the earth's outer shell. Of course, it wasn't the island we see today, but part of a broad, deep oceanic lava plain that supported several chains of volcanic islands. Geologists call these chains **island arcs** because, in plan view, the volcanoes are grouped in curved or arcuate clusters, similar to the present-day Aleutians. These volcanoes were of the explosive type like Mt. Baker and Mt. St. Helens. Dense clouds of hot gas and ash, like those that destroyed Pompeii and Herculaneum almost two thousand years ago, spread outwards from the volcanoes and were deposited on the deep ocean floor as blankets of ash, called **tuff**, along with blocks of lava torn from the vent walls and **volcanic sandstones** eroded from the slopes of the cones. This process continued intermittently for almost twenty million years and resulted in the accumulation of what is called the *Sicker* ***Group*** (of **formations**), named after Mt. Sicker north of Duncan.

The Carboniferous and Permian Periods

About 360 million years ago things quieted down. The ceaseless action of wind and waves had eroded the volcanic island arc to a broad, smooth submarine plateau upon which colonies of marine animals flourished in warm, clear seas. These animals, such as crinoids, brachiopods and bryozoa, contributed their shells of calcium carbonate to the accumulating debris that ultimately became the prominent grey, cliff-forming **limestone** of the *Mt. Mark Formation,* one of three formations of the *Buttle Lake Group* (see photos on pages 23 and 24), which can be seen on Mt. Mark, west of Qualicum. By the close of the Paleozoic Era, some 245 million years ago, what was to become Vancouver Island consisted of layer upon layer of volcanic rocks and limestone that accumulated in a setting not unlike that of modern Indonesia.

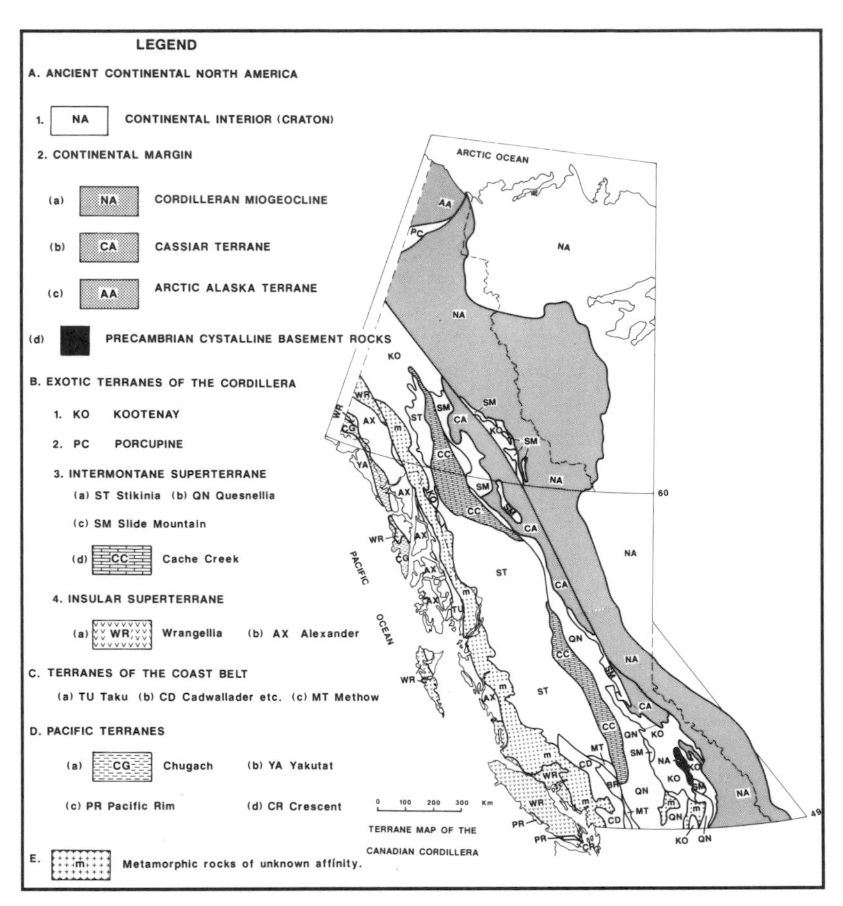

The Canadian Cordillera is made up of a collage of exotic fragments of the earth's crust, called terranes, which collided with the ancient edge of ancestral North America (NA, CA), beginning about 170 million years ago. These terranes, most of which originated as volcanic islands or pieces of sea floor, are given names such as appear in the legend. Prior to collision, several of these terranes amalgamated into giant superterranes, which then accreted to the continent. The Intermontane Superterrane, consisting of Stikinia, Quesnellia, Cache Creek and Slide Mountain, was the first to collide during Middle Jurassic time. The Insular Superterrane, made up of Wrangellia and the Alexander Terrane, is thought to have docked by about 100 million years ago. Since then, smaller terranes have continued to accrete to the continent, including the Pacific Rim and Crescent terranes which form southernmost Vancouver Island.

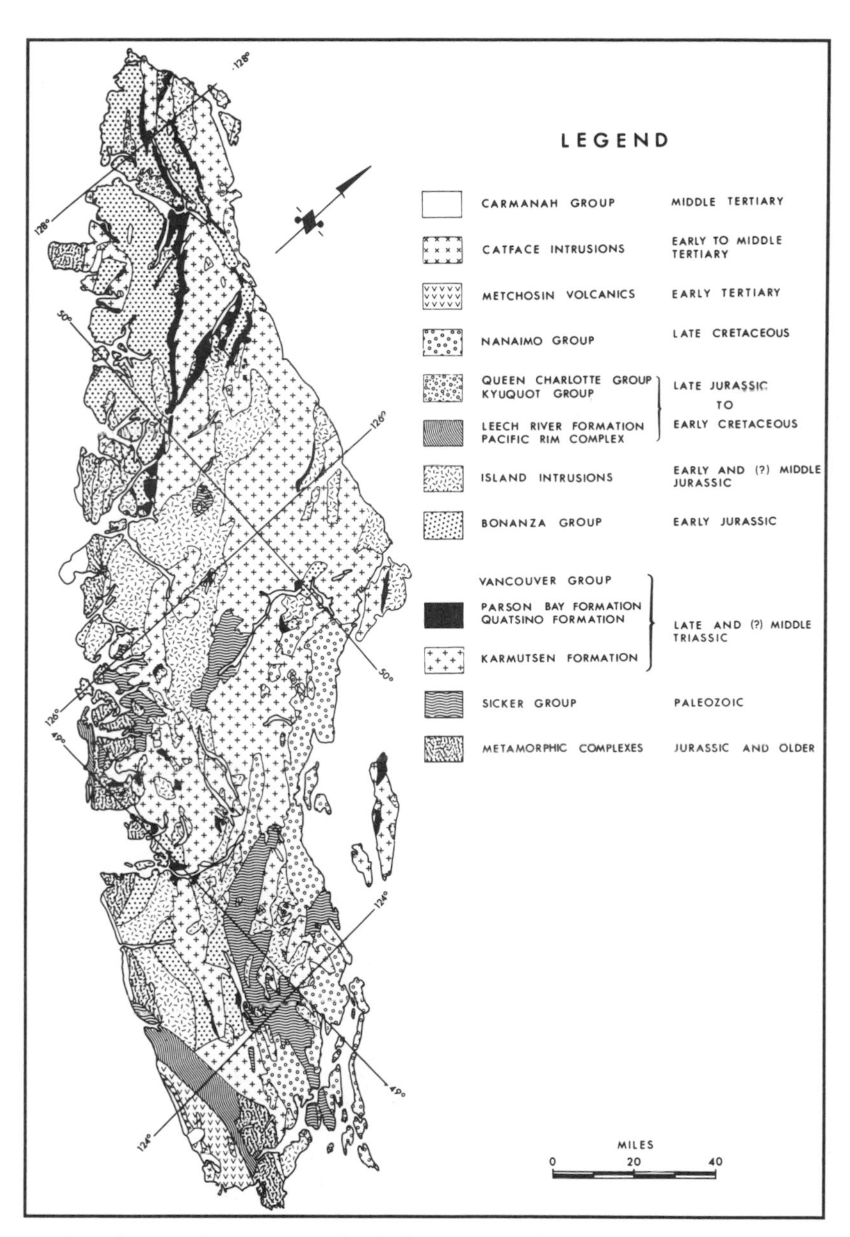

Geological map of Vancouver Island (from J.E. Muller).

Angular blocks and fragments of lava lie in a mixture of ash and dust in this outcrop in the middle part of the Sicker Group near Port Alberni.

THE MESOZOIC ERA

The Triassic Period

The first ten to twelve million years of the **Triassic Period** of the **Mesozoic Era** is a blank. There are no rocks on Vancouver Island that are known to have formed at that time; perhaps it was a time when nothing happened. Perhaps much happened, but the record was lost through erosion. In any event, things got going again about 230 million years ago in the late Triassic Period, when the remains of the old Paleozoic submarine plateau and underlying oceanic crust were split apart. From deep below the crust, masses of lava squeezed up through the fractures and, like blobs of toothpaste, spread outward over the eroded surface of the extinct volcanoes and limestone. The lavas of this phase are chemically different from those of the *Sicker Group*; instead of being shades of green, they are black, are called **basalt** and form the *Karmutsen Formation*, the thickest and most widespread rock formation on Vancouver Island and the oldest formation of the *Vancouver Group*. A drive along the east shore of Buttle Lake near Campbell River allows you to view these layers of lava where they are exposed in mountain faces on the west side of the lake. The lower or older part is composed of **pillow basalt**, a kind of lava that, when extruded beneath the sea, solidifies in pillow-like masses. As the lava pile continued to build upward towards the surface of the sea, its character changed such that

EON	ERA	PERIOD	AGE IN MILLIONS OF YEARS
PHANEROZOIC	CENOZOIC	QUATERNARY	2.0
		TERTIARY	66
	MESOZOIC	CRETACEOUS	135
		JURASSIC	208
		TRIASSIC	245
	PALEOZOIC	PERMIAN	286
		CARBONIFEROUS	360
		DEVONIAN	408
		SILURIAN	438
		ORDOVICIAN	505
		CAMBRIAN	590
PRECAMBRIAN	PROTEROZOIC	UPPER	900
		MIDDLE	1600
		LOWER	2500
	ARCHAEAN		4500

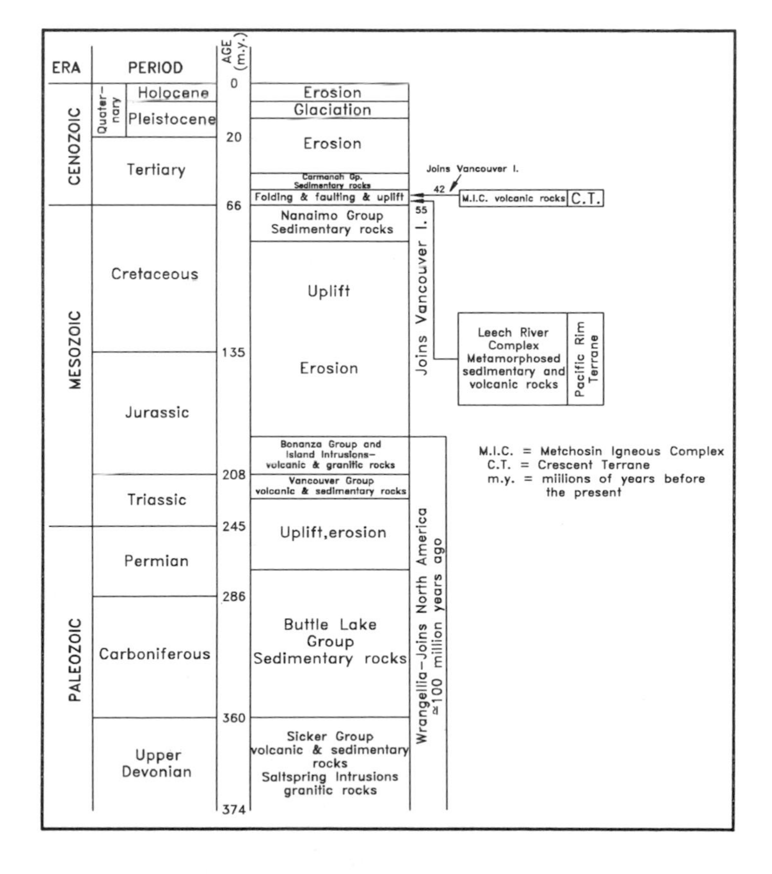

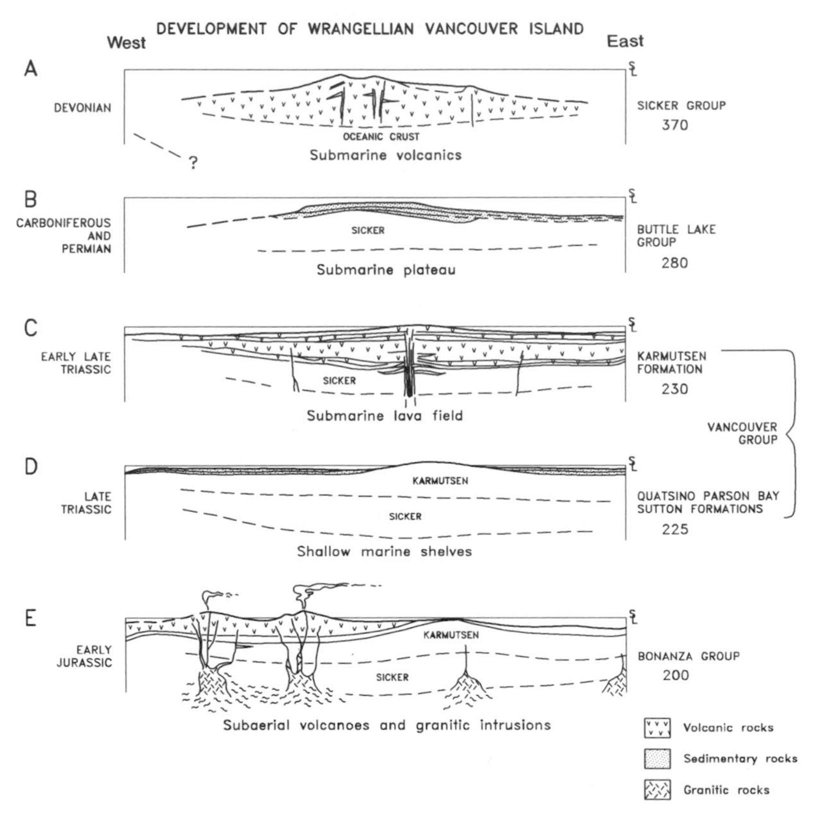

The development of Wrangellian Vancouver Island is here schematically shown by a series of cross-sections, each drawn across the island from approximately Barkley Sound on the west to Parksville on the east. From top to bottom the cross-sections depict the sequential accumulation of volcanic and sedimentary strata and the intrusion of granitic rocks as described in the text. A descriptive title depicting the overall environment of accumulation is given beneath each cross-section.

Upper Paleozoic sedimentary strata of the Buttle Lake Group and Upper Triassic volcanic rocks of the Karmutsen Formation respectively, from the lower and upper cliffs of Mt. Mark at the northwest end of Horne Lake.

the lava flows spread outward in thick layers that extended for great distances across the old surface of the Paleozoic volcanic arc. Layer after layer accumulated until ultimately, when volcanism ceased some five million years after it had begun, the surface of the lava field was close to sea level. There are many localities on Vancouver Island where you can see exposures of the *Karmutsen Formation* (Locality 5, and photo page 26), but few are quite so impressive as those displayed by the castellated cliffs of Mt. Arrowsmith near Port Alberni (see photo on page 12).

Following the end of Triassic volcanism, corals and clams formed colonies on the submerged *Karmutsen* lava plateau; their shell debris, moved by waves and currents, accumulated in isolated depressions on the old lava surface, and, following burial by younger sediments, was converted to limestone of the *Quatsino Formation*. In other places, the *Karmutsen* lavas, and perhaps parts of the Paleozoic *Sicker Group*, were being eroded to gravel, sand, silt and clay, which were carried into deeper water to accumulate as the *Parson Bay Formation*. At a few localities coral reefs flourished, and these became the *Sutton Formation*, the youngest of the *Vancouver Group*, an example of which can be from a boat on the south shore of Cowichan Lake.

The Jurassic Period

At the beginning of the **Jurassic Period**, some two hundred million years ago, arc volcanism again returned to *Wrangellia*. Unlike the

Fossiliferous limestone (light) and thin layers of chert (dark) form the Mt. Mark Formation of the Buttle Lake Group on Mt. Mark near Horne Lake.

period of *Sicker* marine volcanism, which ended one hundred million years earlier, the volcanic rocks of the *Bonanza Group* were formed from lavas that mostly were ejected on land. At the same time, deep beneath the surface, the parent **magma** slowly cooled and solidified into large, irregular masses of granitic rock. The great heat and pressure accompanying the injection of these magmas caused the older rocks to become locally metamorphosed such that the deeper parts of the *Sicker Group* were changed into **gneiss**. More than 150 million years later, these granitic rocks and gneisses were thrust upward, the former ultimately to become exposed as **plutons** of the *Island Intrusions* in the high peaks of the Vancouver Island Ranges and the Saanich Peninsula (Localities 6 & 7, and photo page 27), and the latter as the dark, angular outcrops of *Wark* and *Colquitz gneiss* in the rock gardens, parks and road cuts of Victoria (Localities 1, 2 & 3).

The close of Jurassic volcanism ended the construction of *Wrangellian* Vancouver Island. Subsequent events, such as its accretion to continental North America, the accretion of other, smaller pieces of crust to the island, accumulation of sedimentary strata on top of *Wrangellia* and the profound effects of glaciation all left their modifying signatures on the overall architecture of Vancouver Island. To explain the nature and causes of subsequent events, a short interruption in the story's continuity seems necessary.

A Brief Digression

Let us now digress for a moment. Earlier we said that continents move. Indeed, the familiar theories of **sea-floor spreading** and **plate tectonics** are the modern expressions of an old idea known as **continental drift**. The two theories describe how, in the oceans, the crust that forms the floor of the oceans is continuously being created along the global **mid-ocean ridge** system and how older crust is consumed in what are called **subduction zones** at the edges of the several tectonic plates that make up the earth's outer shell. The subduction of crust occurs along linear, narrow **submarine trenches** at which the volcanic rocks of the sea floor and their sediment cover are bent downwards beneath the continent and descend to depths at which they melt. Some of this molten material rises upward through the overlying **continental crust** to reappear at the surface as volcanoes such as Mount St. Helens, Mt. Baker and the volcanoes of the Aleutian Islands. This process is not new. It has been going on for hundreds of millions of years. In fact, it is believed that the volcanic rocks of the *Sicker* and *Bonanza groups* are the products of subduction processes that occurred long ago along the edges of tectonic plates.

The illustration on page 28 shows sea-floor spreading and subduction as they occur off the west coast of Vancouver Island today. Approximately 200 km west of Tofino, and covered by some 2 to 3 km of water, is the *Juan de Fuca Ridge*. Along the axis of the ridge, molten rock, or magma, is being injected from beneath the crust into fracture systems that allow it to rise toward the ocean floor. Upon meeting the cold seawater the lava cools and forms vertical sheets, called **dykes**, along the axial fractures. Some of the magma flows out onto the sea floor to form pillow basalt. With each succeeding injection, the walls of the ridge move apart, allowing more material to accumulate, thus creating the *Pacific Plate* to the west and the *Juan de Fuca Plate* to the east. The rate of spreading is estimated to be about four centimetres per year, which is roughly equivalent to the speed at which your fingernails grow. As the process continues, the newly created oceanic crust of the *Juan de Fuca Plate* cools, thickens and becomes deeper as it moves towards the edge of the continent.

Meanwhile, back in the Atlantic Ocean, new sea floor is also being generated along the *Mid-Atlantic Ridge* at a rate of about two centimetres per year. The North American continent is being carried like a passenger on a conveyor belt away from this ridge and towards the Pacific. Because the rocks of the continent are comparatively light and thick, they "float" higher than does the thinner and heavier crust of the ocean. At the point of confrontation, where one of the two converging plates has to give way, it is the lighter continent that overrides and causes subduction of the sea floor. Thus, as the comparatively young oceanic crust of the *Juan de Fuca Plate* meets the westerly moving edge of the continent, the denser oceanic crust is overridden by the North American Plate. It

Broken and fragmented pillow lava of the Karmutsen Formation forms the cataract at Englishman River Falls Provincial Park.

is through these processes of sea-floor spreading and subduction, whereby the earth's plates are constantly moving as new crust is created and old crust destroyed, that pieces of crust are moved from one place to another. As the processes continue, things such as volcanoes and coral-limestone reefs, being carried on the oceanic crust, are ultimately welded to the continent as their supporting sea-floor rocks are subducted beneath it. It is by these means that *Wrangellia* was transported and became attached to the ancient edge of North America.

How do we know that the rocks of Vancouver Island have moved about from place to place throughout geologic time? How did they get here? And when? The answer to these questions lies partly in the study of the earth's ancient **magnetic fields** as revealed by the magnetic properties of rocks. When molten magma containing iron-bearing minerals solidifies into rock, its magnetic particles become aligned parallel to the earth's magnetic field. For example, if a lava cooled and solidified today on the main street of Ladysmith on Vancouver Island, a sample of the rocks, when placed in an instrument called a **magnetometer**, would reveal that the **inclination** of its magnetic particles is parallel to the direction of the earth's magnetic field approximately at the forty-ninth parallel of latitude, the latitude of Ladysmith. The magnetic components of a lava formed at the equator would show zero inclination, whereas one that cooled near the north magnetic pole

The Little Qualicum River has cut its gorge into grandiorite of the Island Intrusions at Little Qualicum Falls Provincial Park.

would show vertical inclination. Thus, if we know the age of a rock and can measure the inclination of its magnetization, we can then determine the latitude where the rock formed. We must assume, of course, that the earth's magnetic poles have always been close to its rotational, or geographic, poles. Once we have determined the latitude where a rock formed and its age, we can then calculate the average rate at which it moved to its present location.

During that drive along the east shore of Buttle Lake we talked about earlier, you might have noticed that on the side of the road, several outcrops of black volcanic rocks of the *Karmutsen Formation* have holes in them. These are not blast holes, but small core holes drilled by Ted Irving, an internationally famous geoscientist of the Geological Survey of Canada (retired) who lives near Victoria. A short time before Ted's work in the area, Tim Tozer of the Survey had discovered tropical varieties of fossil organisms in limestone near the top of the *Karmutsen Formation*, suggesting that these and older rocks on Vancouver Island had not always been part of North America but had come from much farther south. Ted wanted to test this idea by determining the latitude where the rocks formed. He and Ray Yole, a colleague from Carleton University, collected several drill-core samples of the *Karmutsen Formation* at many places along the shore of Buttle Lake, then took their samples back to the laboratory, where they were subjected to detailed laboratory analyses. Their results indicated that, about 230 million years ago, the *Karmutsen Formation* of Vancouver Island (*Wrangellia*) was formed probably far to the south. Since that time *Wrangellia*, along with the *Alexander Terrane* (together forming the *Insular*

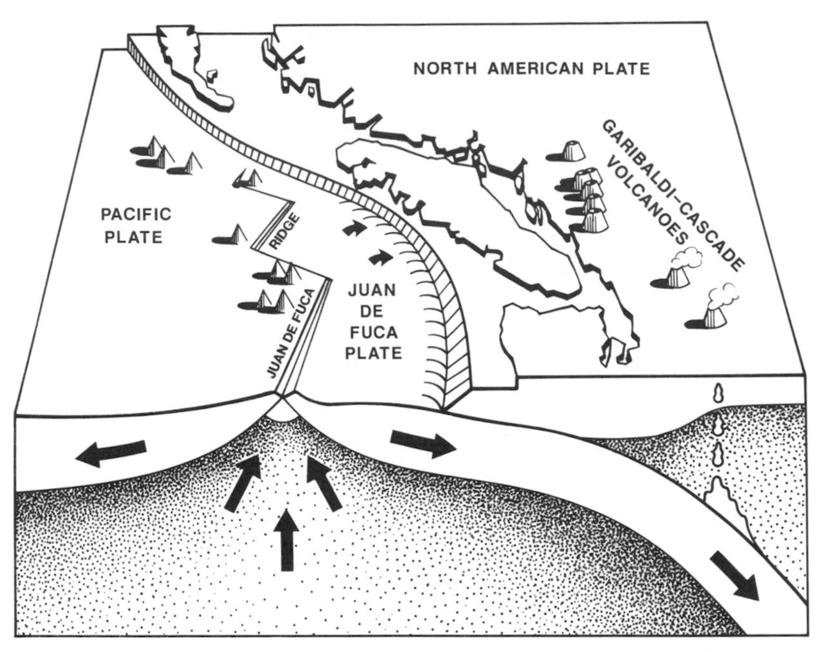

This three-dimensional drawing portrays the plate tectonic situation off the coast of Vancouver Island and the southern Queen Charlotte Islands. Molten material from deep below the ocean floor rises upward (upward arrows) and is injected into the axis of the Juan de Fuca Ridge, where it solidifies as igneous dykes which form much of the oceanic crust. With each succeeding injection, the oceanic crust grows or spreads outwards (horizontal arrows) such that the plates grow symmetrically away from the ridge to form the Pacific Plate on the west (left) and the Juan de Fuca Plate on the east (right). Where the latter meets the westward-moving continent, it descends, or is subducted back into the mantle beneath the continental crust of North America (downward arrow). At a depth of between 150 and 200 km, the plate melts, from where molten material again rises upwards to the surface to reappear as the Garibaldi-Cascade volcanoes, such as Mt. Garibaldi, Mt. Baker, Mt. St. Helens and many others.

Superterrane), is believed to have moved northward and to have become accreted to North America by middle Cretaceous time, some one hundred million years ago.

Western British Columbia was assembled in piecemeal fashion. Through the mechanism of sea-floor spreading, various chunks of crust, or terranes, were carried northward from southern latitudes, and, after amalgamating with other pieces into superterranes, they successively collided with the western edge of the ancient continent throughout the past 170 million years. *Wrangellia*, the westernmost terrane, is named from the Wrangell Mountains of southeast Alaska and includes the Queen Charlotte Islands, most of Vancouver Island and parts of the Coast Mountains. Other fragments in the collage of British Columbia are called *Stikinia*, *Quesnellia* and the *Cache Creek* and *Slide Mountain terranes*; all of these are foreign fragments that were added to the western edge of North America throughout the past 170 million years (see diagram on page 18).

The Cretaceous Period

So now we return to our story and pick up *Wrangellia* on its northbound journey. By Cretaceous time the part of *Wrangellia* that ultimately became Vancouver Island consisted of Paleozoic volcanic rocks and limestone (*Sicker Group* and *Buttle Lake Group*), Upper Triassic oceanic lava and sediments (*Vancouver Group*), and Lower Jurassic volcanics that exploded outwards over the ancient lava plateau (*Bonanza Group*). By this time *Wrangellia* had travelled northward to about 1,000 km south of its present position. The remainder of the trip took about eighty-three million years. Close to a hundred million years ago, during the middle part of the Cretaceous Period, when Africa was separating from South America to create the south Atlantic Ocean, *Wrangellia* crashed into North America.[1]

It was during the times of collision that much of the structure and overall character of British Columbia developed. Although they took several million years to complete, the collisions resulted in crushing, folding and faulting in the colliding terranes. On Vancouver Island, *Wrangellia* was compressed and buckled upward to form two mountainous regions called the *Cowichan Anticlinorium* and *Buttle Lake Anticlinorium*, which extended from Saltspring Island through Cowichan Lake to Horn and Buttle lakes. As they were uplifted these mountains were exposed to the forces of erosion. Boulders, cobbles, pebbles, sand, silt and clay were washed from the mountains into nearby basins. On their western slopes several formations assigned to the *Coal Harbour Group* accumulated in the Quatsino Sound area. Somewhat later, about eighty-five million years ago, a broad basin formed along the east coast of the island and in the Strait of Georgia region, into which streams carried sand, gravel and mud eroded from the adjacent mountains. The edges of the basin were low, swampy areas, rich in forest vegetation that became transformed into coal. These sediments and the coal are called the *Nanaimo*

---

[1] It should be noted that there are other arguments pertaining to the latitudes of origin and times of accretion of these terranes with one another and with North America. Readers interested in these are referred to "Additional Sources" at the end of this guidebook.

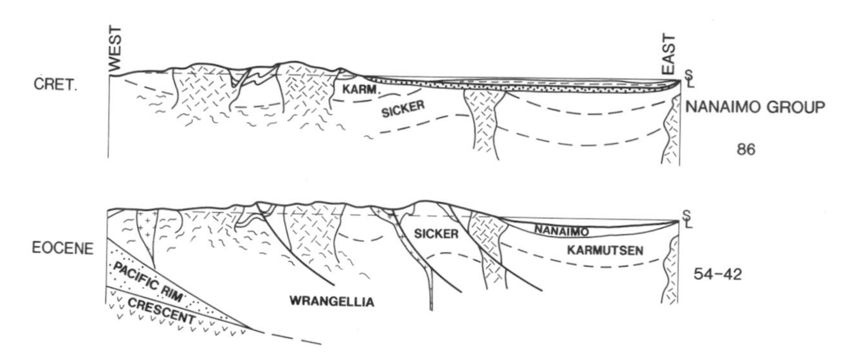

As a consequence of the accretion of Wrangellia to North America, the core of Vancouver Island was uplifted. Sediments eroded from these elevated regions were transported eastwards to fill a sedimentary basin that underlies the east coast of the island and the Strait of Georgia (Nanaimo Group, upper cross-section). During the Eocene Epoch of the Tertiary Period, the Pacific Rim and Crescent terranes were sequentially added to the island. Forces arising from these collisions caused further uplift of the island and fracturing by thrust faults (lower cross-section).

Group, which underlies the coastal plain (*Nanaimo Lowland*) of the island from south of Campbell River to the Saanich Peninsula (Locality 10) and which also forms most of the rocks of the Gulf Islands (Locality 20, and photo page 32).

At the northern end of the island, near the communities of Port McNeill and Port Hardy, rocks assigned to the *Nanaimo Group* form low, coastal outcrops of sandstone and shale. Other areas that display parts of the group are the ski slopes of Mt. Washington, the Cowichan Valley and Duncan area, the Alberni Valley and the northernmost San Juan Islands.

THE CENOZOIC ERA

The Tertiary Period

But things were not over yet. Two important pieces of Vancouver Island were yet to be added. Along the west coast at Pacific Rim National Park and in an area extending easterly from Port Renfrew, near the entrance of the Strait of Juan de Fuca, to Victoria, are rocks assigned respectively to the *Pacific Rim* and *Leech River complexes*, both of which are included in the **Pacific Rim Terrane**. At their original location, perhaps in the vicinity of the modern San Juan Islands, these rocks formed from large submarine landslides and other continental slope deposits. They moved from the southeast to their present location along the *Westcoast*

Fault and *San Juan – Survey Mountain Fault*, about fifty-five million years ago, when several volcanoes were erupting along the west coast. These rocks are best seen along the coast between Ucluelet and Tofino, in the Leech River country west of Victoria (Locality 16) and along the Malahat north from Goldstream Park to near the turnoff on Aspen Road (Localities 8 & 9).

Shortly thereafter, about forty-two million years ago, a second terrane was added. It is called the ***Crescent Terrane***, which, on Vancouver Island, consists of rocks assigned to the *Metchosin Igneous* **Complex**. The complex consists of ancient sea-floor volcanic rocks upon which live the people of Sooke, Metchosin and Colwood and which form the magnificent Olympic Mountains across the Strait of Juan de Fuca (Localities 11, 12 & 13). The *Crescent Terrane* was shoved under the *Pacific Rim Terrane* along the *Leech River Fault*, the surface trace of which is an arcuate, V-shaped valley that extends easterly from Sombrio Point (Locality 16) on the north shore of the Strait of Juan de Fuca to Esquimalt Lagoon near Victoria. All of this was happening during approximately the same time as the subcontinent of India crashed into the underside of Asia to create the Himalayas.

An important effect of these latter collisions was that they caused the western and southern edges of *Wrangellia*, or Vancouver Island, to be lifted into the air as if by a giant wedge rammed in beneath them. The forces of erosion quickly attacked the elevated landscape, wearing away as much as 10 km of rock and revealing the deeper parts of the island's crust represented by the *Wark* and *Colquitz gneiss complexes*, upon which the city of Victoria is built (Localities 1, 2 & 3). An additional effect of these collisions was the folding and faulting of the *Nanaimo Group* sediments which led to the formation of the Gulf Islands.

On southern Vancouver Island, the Tertiary Period ended with the accumulation of sandstone, conglomerate and shale of the *Carmanah Group*, most of which occurs beneath the **continental shelf** off the west coast. Along the shores of the Strait of Juan de Fuca, however, the *Carmanah Group* is represented by the *Sooke Formation*, consisting of sandstone and lesser conglomerate that accumulated in shallow marine waters (Localities 12, 14, 15 & 17). On northern Vancouver Island, a series of small volcanic cones erupted between 8 and 3.5 million years ago. On Haddington Island near Port McNeill, the lavas of these volcanoes have been quarried for use as building stones (some of which were used in the construction of the Parliament Buildings in Victoria).

Characteristic coastal outcrops of the Upper Cretaceous Nanaimo Group in the Gulf Islands consist of massive conglomerate (top) and interstratified siltstone and shale (bottom).

The Quaternary Period

Introduction

The **Quaternary Period** of the **Cenozoic Era** began about two million years ago and is divided into two epochs, the **Pleistocene** and the **Holocene**, or Recent. The Pleistocene is commonly known as the Ice Age, and was characterized by worldwide climatic fluctuations and glaciations throughout the northern hemisphere.

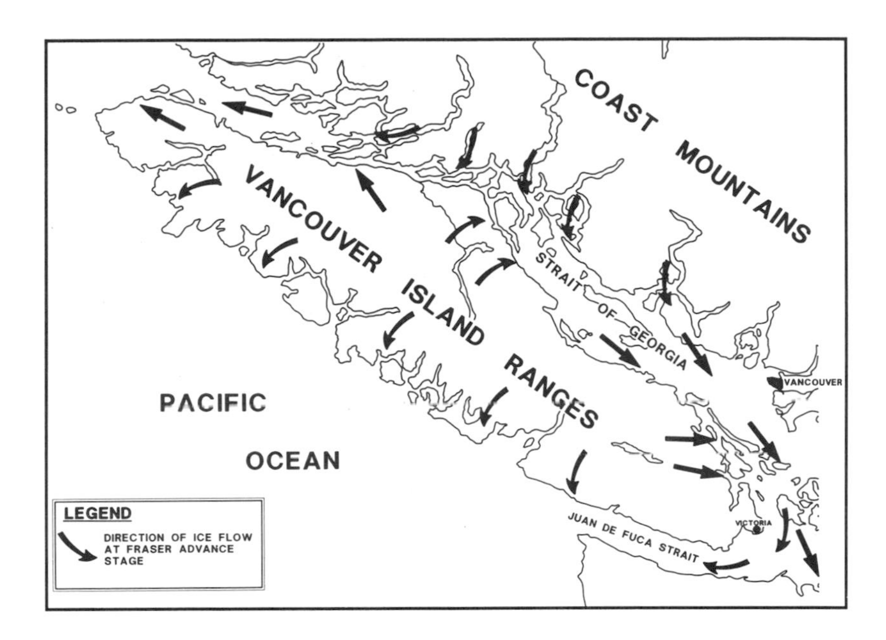

These glaciations were not continuous, but were separated by long periods when the climate was as warm or even warmer than at present. The Pleistocene ended about ten thousand years ago when ice from the last glaciation melted and the present pattern of climate was established. Whether or not the Recent Epoch is only a warm interval between glaciations remains to be seen. On Vancouver Island, the events and processes of the Quaternary Period resulted in the deposition of the unconsolidated soils that overlie bedrock and produced many of the physiographic features of the island's landscape.

Glacial Record of Vancouver Island

Vancouver Island was subjected to at least three glaciations separated by two nonglacial intervals. At most places, traces of the earlier glacial and nonglacial events have been scoured away or buried by sediments from younger glaciations. Only features related to the ***Fraser Glaciation*** and postglacial time are widespread and well exposed.

The deterioration of climate marking the onset of *Fraser Glaciation* began about twenty-nine thousand years ago. Ice began to accumulate on the mountains of Vancouver Island and the mainland to form alpine glaciers similar to those which can be seen today only on the highest peaks of the island. As the climate became cooler and wetter, these alpine glaciers expanded and coalesced to form valley glaciers, which filled major valleys such as the Cowichan Valley. Similar valley glaciers formed on the

EPOCH		AGE		Local names of events for Vancouver Island
Yrs.	Recent	Holocene		Post Glacial
10000 20000	Pleistocene	Wisconsinan	Late	Fraser Glaciation
30000 40000			Middle	Olympic Interglacial Interval
50000 ≈ 80000			Early	Semiahmoo Glaciation

coastal mountains of the mainland and joined with the glaciers from Vancouver Island to fill the *Coastal Depression*, which included Hecate Strait, Queen Charlotte Sound, the Strait of Georgia and surrounding lowlands. At the south end of Vancouver Island, southerly flowing ice divided into two lobes. One lobe flowed westerly and northwesterly along the Strait of Juan de Fuca to the Pacific Ocean, while the other moved south along Puget Sound, past the location of Seattle, and terminated in the vicinity of Tacoma.

Further accumulation of glacial ice ultimately filled the *Coastal Depression* such that ice flowed across Vancouver Island, probably to terminate in an ice shelf off the west coast. The maximum advance of the *Fraser Glaciation* was reached about fifteen thousand years ago, after which the climate began to warm up. Glaciers had melted and retreated into the mountains close to their present positions and elevations by about ten thousand year ago.

The directions of ice movement during the advance and at the time of maximum extent of *Fraser Glaciation* are shown in the illustration on page 33. At its maximum extent, the weight of the ice depressed Vancouver Island by at least 150 m and possibly by as much as 300 m. At the same time, the amount of water that was removed from the sea and tied up in the continental glaciers caused the worldwide sea level to fall by at least 150 m. Consequently, when examining old glacial beaches and shorelines that today are much higher than current sea level, it is necessary to consider these two effects. In most cases, the degree of depression due to the weight of ice was considerably greater than the amount by which sea level was lowered. When the weight of the ice was removed, the land rebounded back to its

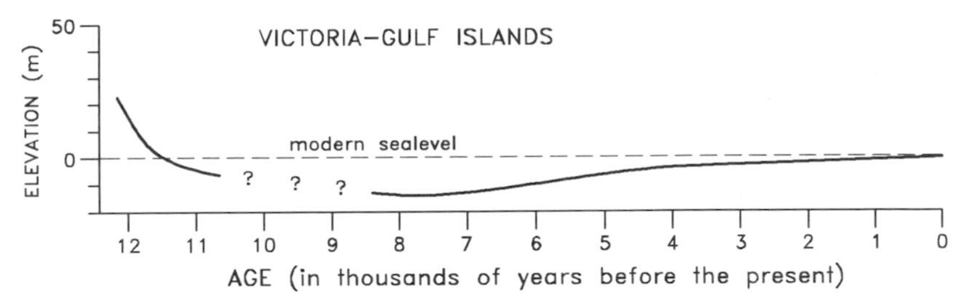

The graph shows the local elevation of sea level relative to modern sea level for the past twelve thousand years, following the retreat of Fraser Glaciation.

former elevation. In this way, former late glacial shorelines are today seen at higher elevations than those of today.

When the ice retreated from Vancouver Island about thirteen thousand years ago, the shoreline along Georgia Strait was 60 to 120 m higher than at present. However, because the land was rising more quickly than the sea, for about a thousand years the relative shoreline actually fell until it was lower than the present coastline. When rebound stopped, the oceans continued to rise slowly until about five thousand years ago, when sea level reached its present position.

Evidence of Fraser Glaciation

Fraser and earlier glaciations modified the landscape of Vancouver Island by erosion and deposition. The most widespread evidence of glacial erosion is the almost complete absence of soil on hilltops, mountains and ridges. Glaciers moving over the surface scraped away all soft weathered rock, particularly from higher elevations, the result being that prominent rock outcrops on Vancouver Island commonly have fresh unweathered surfaces. Moreover, the rock fragments embedded in the base of the ice scratched and scoured the rock surface, leaving conspicuous grooves and scratches on their surfaces (Locality 4). In some instances the ice removed large masses of bedrock, particularly on the downstream side of rocky prominences (Locality 3). This glacial erosion produced a multitude of minor features, including **grooves**, **striations**, **roche moutonnées** and **drumlins**. In alpine areas glacial erosion resulted in the formation of **cirques**, and at lower elevations, the carving of **U-shaped valleys** and fiords. Deposits resulting from waning *Fraser Glaciation* include those which accumulated at and beneath the melting ice front, as well as materials carried by meltwater and deposited on land or in the sea far from the ice front. Glacial **till** is a dense, unsorted mixture of sand, gravel, silt and clay deposited beneath the base of a moving glacier. **Erratic**

Along the north shore of Cadboro Bay, the Vashon glacial till overlies laminated outwash silt of the Quadra Formation. The till was deposited beneath the base of a moving ice sheet during Fraser Glaciation.

boulders, and drop stones from floating ice, are other examples of materials that accumulate from melting glaciers (Locality 8).

As the Fraser glaciers advanced southward along the Strait of Georgia and surrounding lowlands, a widespread deposit of outwash sand and gravel accumulated in front of the advancing ice. These outwash deposits are known collectively as the *Quadra Formation* (Locality 18). The Quadra sands and gravels are exposed at many localities along the east coast of Vancouver Island, as well as on Quadra Island and James and Sidney islands near Victoria. In Victoria, bones of the Imperial Mammoth, muskox, horse, bison and mastodon have been discovered in gravel pits excavated in the *Quadra Formation.* These animals may have crossed over to Vancouver Island on the large floodplain that formed in front of the ice as it advanced down the Strait of Georgia.

During the time when Vancouver Island was completely covered by ice, the only glacial debris deposited was till, which accumulated beneath the base of the ice sheet and on top of the older outwash materials of the *Quadra Formation.* This material, called the *Vashon Till* (Localities 4 & 18, and photo page 37), is widespread in road cuts and sea cliffs, and common in gravel pits throughout the Greater Victoria area. As the *Fraser Glaciation* waned, vast amounts of water were released from the melting ice to form streams and rivers that washed sand, gravel, silt and clay away from the glaciers toward

A close-up view of the Vashon Till shows its wide range of particle sizes.

the sea. At that time the shoreline was relatively higher than at present, and at the mouths of most rivers outwash deltas, such as the *Colwood Delta* near Victoria (Locality 19), were formed. Far from the river mouths, silts and clays settled onto the sea floor, and as the land rose and shoreline fell, these silts and clays were exposed on lowlands of Vancouver Island, where they form much of our modern agricultural soils.

At the beginning of the Recent Epoch, about ten thousand years ago, the climate of Vancouver Island had warmed. About 6,600 years ago, cooler and wetter conditions began to become established, thus developing the climate that prevails today. Sea level at the beginning of the Recent Epoch was as much as 10 m lower than it is today. As sea level rose, and the shoreline stabilized at its present level, storm waves cut cliffs in the unconsolidated coastal deposits, reworking them into sand and gravel beaches. Where there is an abundant supply of material and strong longshore currents, these materials have been swept parallel to shore to form spits and offshore bars.

During the past ten thousand years, streams and rivers on Vancouver Island have cut down through the Pleistocene deposits and washed sand and gravel to the sea to build deltas, which are found at the mouth of all major rivers. In the process, the rivers have cut numerous terraces. Valley slopes have been modified by landslides, the scars of which are evident on steep valley slopes in mountainous areas.

STRUCTURE OF VANCOUVER ISLAND

The geological architecture of Vancouver Island is a product of both the types of rocks that comprise the earth's crust in this region and the tectonic forces that, throughout geological time, have deformed them into a wide variety of structures. Indeed, apart from the effects of glaciation, the dominant controls on landforms and physiography are rock type and **structure**. By structure we mean the many kinds of **faults** and **folds**, their orientation, magnitude and complexity. Taken altogether, they determine the **structural style** of a region.

The structural style of Vancouver Island is displayed by a dominant northwesterly trending grain or fabric. **Igneous**, **sedimentary** and **metamorphic** rocks of different ages and modes of formation occur as northerly or northwesterly trending narrow to broad zones; the boundaries between them commonly are important faults. Within these zones are many other faults of variable significance, which, together with folds, locally disrupt the dominant northerly to northwesterly trending grain. In southernmost Vancouver Island, major westerly trending faults partition the crust into two narrow zones, each of which consists of rocks that were accreted to the island very late in its development, between about fifty-five and forty-two million years ago. The linear, northwesterly trending pattern of the Gulf Islands is a reflection of folds and faults that are believed to have formed when these two fragments of crust were added to the island.

Along its length, the island's highest mountains have been carved from granitic rocks, which form large masses, or plutons, and which have northerly orientations. For the most part they are surrounded by thick accumulations of lava into which the plutons were intruded about two hundred million years ago. Because granitic rocks are comparatively resistant to the forces of erosion, they commonly form prominent hills and peaks, such as Mt. Newton (Locality 6) and Bear Hill, north of Victoria.

Two important structures on the island are the *Cowichan Anticlinorium* and the *Buttle Lake Anticlinorium*. These are large, northwesterly trending upfolds of the crust wherein the oldest rocks on the island have been exposed by erosion. The *Cowichan Anticlinorium* extends in an arcuate, northwesterly trend from

Saltspring Island to the north side of the Alberni Valley, whereas the smaller *Buttle Lake Anticlinorium* occurs in the central part of the island near Buttle Lake. In each of these structures, many smaller **anticlines** (up-folds) and **synclines** (down-folds) add complexity to outcrop patterns, as do large, bordering faults. As noted in the previous section, these large structures are thought to have formed at the time of collision of *Wrangellia* with North America. Erosion of these uplifted structures resulted in the accumulation of the *Nanaimo Group.*

Because faults are such important features of the geology of Vancouver Island, the following discussion focuses on some of their more important representatives on the southern part of the island.

On Vancouver Island and throughout the Gulf Islands, **thrust faults** and **normal faults** appear to be the most common kinds of faults. Thrust faults are gently inclined fractures separating blocks of rock of which the block above the fault has moved upward and over the block beneath. Normal faults generally are steep fractures separating blocks of rock of which the block above the fault has been dropped relatively downward beside the block beneath. In many areas of the island, much of the surface is mantled by soil, glacial drift and dense tree cover, making fault recognition difficult. Of the known larger and more important faults, the fault lines or **surface traces** of most are aligned in a northwesterly direction, although faults with other trends have been mapped, particularly on central and northern Vancouver Island.

The main times of movement of the principal faults on southern Vancouver Island were probably between about fifty-five and forty million years ago, the interval when the *Pacific Rim* and *Crescent terranes* accreted to *Wrangellia.* No faults are known to be active today, nor are there any indications of activity since the beginnings of settlement in this region. No earthquakes have been identified as occurring on a specific fault; however, precise location of earthquake **epicentres** and their **focal depths** has not always been possible. Improved earthquake detection capability on Vancouver Island and systematic surveys of an epicentre region may one day be successful in identifying renewed activity along these faults.

To fully appreciate the following descriptions of some of the principal faults on southern Vancouver Island, readers should avail themselves of the 1:50,000 scale topographic maps and the 1:100,000 scale geological map listed on page 3. For a more detailed discussion of fault types, the reader is referred to the glossary.

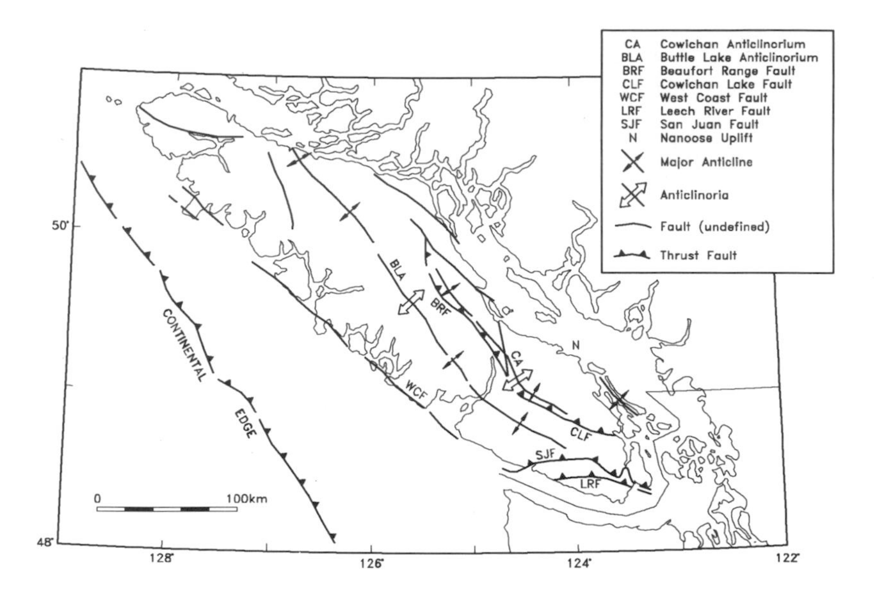

The Leech River Fault

The *Leech River Fault* is one of the most prominent faults in this region and is easily recognizable on topographic maps. The fault occurs on southern Vancouver Island and beneath the continental shelf where it separates volcanic sea-floor rocks of the *Crescent Terrane* from sedimentary and volcanic rocks of the *Pacific Rim Terrane.* On land the fault extends westerly for some 153 km from Esquimalt Lagoon in Greater Victoria to Sombrio Point on the north shore of the Strait of Juan de Fuca (Locality 16). For much of its length the fault occupies a narrow, steep-sided, prominent valley within which are found Leech River, Bear, Wye and Loss creeks (see photo on page 41) and Bear and Diversion reservoirs. It is a northward-dipping structure, the inclination of which is about 60° at the surface, decreasing to between 35° and 45° at a depth of about 3 km.

The San Juan – Survey Mountain and Westcoast Fault System

The *San Juan – Survey Mountain Fault* on southern Vancouver Island separates the *Pacific Rim Terrane* from *Wrangellia.* The northward-inclined *San Juan Fault* extends from close to the north shore of Port San Juan in an arcuate easterly trend along the San Juan River to close to its confluence with Clapp Creek, where it continues northeasterly, with poor topographic expression, toward Cobble Hill and Satellite Channel. A good exposure of the *San Juan Fault* occurs on the north side of the junction of Port Renfrew Road and the South San Juan main logging road,

The valley of Loss Creek, west of Victoria, marks the surface trace of the Leech River Fault, along which rocks of the Crescent Terrane (left) were emplaced beneath those of the Pacific Rim Terrane (right) about forty-two million years ago.

close to where the former crosses Clapp Creek. There the rocks of both *Wrangellia* and the *Pacific Rim Terrane* are intensely shattered in a zone approximately 2 km wide. Large blocks of dark volcanic rocks (*Wrangellia*), a few several metres in size, are glued together in a matrix of mushy slate (*Pacific Rim Terrane*).

A short distance north of Weeks Lake, the *San Juan Fault* is intersected by the southeasterly trending *Survey Mountain Fault*. In the area surrounding the point of intersection, the rocks of *Wrangellia* and the *Pacific Rim Terrane* have been squeezed, crumpled and broken by many subsidiary smaller faults. From Weeks Lake the *Survey Mountain Fault* trends southeasterly, crossing Cragg Creek on the southeast side of Survey Mountain, the southernmost end of Sooke Lake and passing immediately south of the south end of Goldstream Lake. From there the surface trace of the fault turns abruptly northward to cross Malahat Drive near Aspen Road, where it swings sharply southward again as it crosses the glacial fiord of Finlayson Arm to traverse the west flank of the Gowlland Range and the southwest shoulder of Mt. Finlayson above Goldstream Park (Locality 9). On Mt. Finlayson, metamorphosed igneous rocks of *Wrangellia* form the main cliffs of the mountain, whereas the more easily eroded metamorphosed sedimentary rocks of the *Pacific Rim Terrane* form the lower, gentler slopes. The trace of the fault through the communities of Langford and Colwood is obscured by a thick cover of glacial deposits forming the *Colwood Delta* (Locality 19). The fault reappears at Fort Rodd Hill, from where it continues offshore.

As its name suggests, the *Westcoast Fault* occurs along the west coast of Vancouver Island, where it is discontinuously exposed on several peninsulas and islands from Brooks Peninsula to near Ucluelet at the south end of Pacific Rim National Park. The fault zone consists of two or more branching strands within a wide zone of crushed and shattered rocks. In the region of Pacific Rim National Park, the fault zone forms the boundary between the sedimentary, metamorphic, volcanic and granitic rocks forming the bulk of the island (*Wrangellia*) and sedimentary and volcanic rocks of the *Pacific Rim Terrane*. Based upon geophysical data, it is known that the fault is inclined towards the northeast at about 40° and extends deep into the crust. On the basis of similar geophysical information, the *Westcoast Fault is thought to be the continuation of the San Juan Fault.*

The Cowichan Lake Fault System

The *Cowichan Lake Fault System* can be traced for some 120 km from Cowichan Lake to the Beaufort Range on the eastern flank of the Alberni Valley. Along the southwestern flank of McLaughlin Ridge in the Port Alberni area, the topographic expression of the system is subdued in comparison to that in the Cowichan Lake region, where it occurs near the base of the steep, southwesterly-facing slope of Cowichan Valley.

The system consists of two main faults, the surface traces of which are arcuate, from a trend of about 335° in the Alberni Inlet area to 305° in the southwestern part of the region to near 295° at the west end of Cowichan Lake. The southwestern strand is inclined from near vertical at Limestone Mountain to 65° eastward on the upper Franklin River. The eastern strand is largely inferred from sudden changes in rock type, thus its surface trace and inclination are not directly observed. The western strand is well exposed on the eastern upland slope of Patlicant Mountain, Limestone Mountain, upper Franklin River and above Redbed Creek.

The *Cowichan Lake Fault System* is considered to be a structure involving high angle, reverse displacement in the order of 2,500 m. The system is thought to be linked with the *Beaufort Range Fault System* in the Port Alberni area.

The Beaufort Range Fault System

The *Beaufort Range Fault System*, in the Port Alberni region, can be traced for 27 km from the northwestern corner of the Horne

Looking southeasterly along Cowichan Lake, the Cowichan Lake Fault Zone occurs near the base of the range on the north (left) side of the lake.

Lake area to Rogers Creek south of Highway 4 (Alberni Highway) near Port Alberni. To the northwest it extends at least as far as Oyster River, for a total length of over 70 km. Northwest of Port Alberni the system comprises mainly two faults, which are about 1 km apart. The lower fault occurs near the break in slope between the Alberni Valley and the Beaufort Range. It is not commonly exposed and is identified by abrupt changes in rock type across the fault. The upper strand, located on the mountain slope, can be traced with greater confidence. The upper fault is exposed in numerous narrow creek valleys draining the southwestern flank of the Beaufort Range, below Mt. Joan and Mt. Irwin, and is well exposed on the road cut along the Alberni Highway. Shatter zones vary from 0.2 to 10 m wide and, adjacent to fault surfaces, gouge up to 0.3 m wide is present. In the vicinity of its link with the *Cowichan Lake Fault System*, the *Beaufort Range Fault System* comprises up to five faults, which presumably resulted from the transfer of stress from one system to another.

The surface traces of the faults belonging to the *Beaufort Range Fault System* are moderately sinuous with an average trend of 325°. The inclinations of the faults vary from near vertical to 65° northeastward at the surface; however, based upon geophysical information beneath the surface, the fault system is believed to dip less steeply with increasing depth to an inclination of about 40°.

Paleozoic and Triassic rocks forming the Beaufort Range on the east have been thrust over Cretaceous strata underlying the Alberni Valley to the west along the Beaufort Range Fault Zone. The traces of the two strands of the fault zone occur in the lower and middle slopes of the range.

The Cameron River Fault System

The *Cameron River Fault System* can be traced for over 60 km from the valley of upper Qualicum River, northwest of Horne Lake, southwest to Nanaimo River. Farther to the southwest, the system probably links with the *Fulford Fault* on Saltspring Island for a total length, on land, of 140 km. The trace of its probable continuation through the southern Gulf Islands is unknown.

The system consists of two main faults, which, in its central portion, are about 1 km apart; to the southeast the separation increases to 3 km. The central part contains several short, branching and subsidiary faults. Along most of its trend, the system has prominent topographic expression and controls the course of the Cameron River. Its surface trace is arcuate and varies from a trend of 320° in the north to 285° in the southeast. Inclinations are steep and variable from 70 to 75° northeastward; however, some subsidiary fractures internal to the system are inclined steeply westward. The system is well exposed near Labour Day Lake and on the road along the southwest flank of Mt. Arrowsmith as far as the Arrowsmith ski development.

The type of motion on the *Cameron River Fault System* is uncertain. Although arguments may be expressed for horizontal displacement of up to 8 km, supporting geological evidence is not strong. The steep northeasterly dip of the main faults, which separate younger rocks on the southwest from older strata on the northeast, point to thrust fault movement similar to the *Beaufort Range* and *Cowichan Lake fault systems.*

Faults of the Gulf Islands and Southeastern Vancouver Island

Southeastern Vancouver Island and the Gulf Islands are largely composed of sedimentary rocks deposited during the latter stages of the Cretaceous Period, between eighty-five and sixty-five million years ago. These strata are included in the *Nanaimo Group*, which is divided into nine formations consisting of various combinations of conglomerate, sandstone, siltstone, shale and coal. In the Nanaimo area these strata are horizontal or very gently dipping; however, in the Gulf Islands they have been deformed into a series of northwesterly trending folds and thrust faults (Locality 20). Faults are common on the inner Gulf Islands, such as Saltspring and Pender, whereas folds are dominant on the outer islands, including Saturna, Mayne, Galiano, Valdez, Kuper and Thetis.

The *Fulford Fault* is a northwesterly trending and northeasterly inclined thrust fault along which rocks of Paleozoic age on the northeast side have been thrust over those of Cretaceous age to the southwest. On Saltspring Island its surface trace follows the low valley extending from Fulford Harbour to Burgoyne Bay. Across Sansum Narrows it traverses the community of Maple Bay, from where it continues westward to cross the Island Highway north of Duncan. From there it crosses the ridge along Bonsall Creek, north of Mt. Prevost, to ultimately link up with the *Cameron River Fault System* discussed above. Southeast of Saltspring Island the *Fulford Fault* continues toward Stewart Island in the San Juan Islands, possibly to link up with thrust faults on northern Orcas Island.

South of the *Fulford Fault* is the *Tzuhalem Fault* and the *Chemainus River Fault*, each of which are thrust faults separating older rocks on the northeast side of the faults from younger strata to the southwest. The *Tzuhalem Fault* occurs on the southwesternmost tip of Saltspring Island, from where it extends close to the north shore of Cowichan Bay and along the south flank of Mt. Tzuhalem, thence to probably link up with one of the strands of the *Cowichan Lake Fault System.* The *Chemainus River Fault* occurs largely within the Chemainus River valley and may form a fracture surface linking the *Cameron River* and *Tzuhalem faults.* The *Pender Fault* is one of several northwest-trending faults that

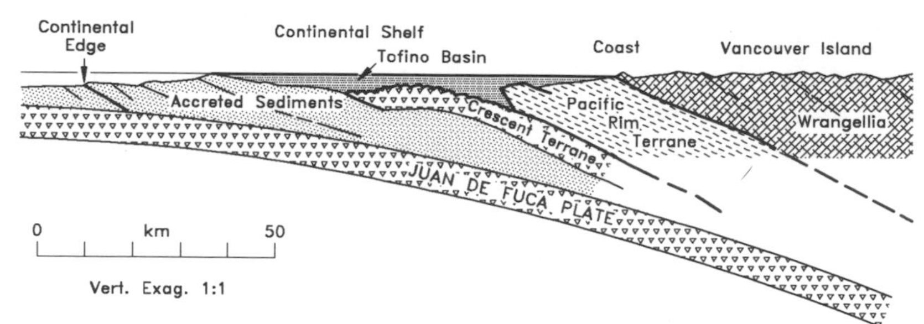

This cross-section, drawn approximately along the latitude of Barkley Sound, shows the structure of the margin of the continent from near the east coast of Vancouver Island to beyond the toe of the continental slope. The Pacific Rim and Crescent terranes successively occur beneath Wrangellia and, in turn, are overlain by sedimentary strata of Tofino Basin, which underlies the continental shelf. Beneath the Crescent Terrane, sediments overlying the subducting Juan de Fuca Plate are being scraped off the oceanic crust and accreted to the overriding North American continental plate.

disrupt the rocks of Pender Island. It extends from near the south shore of Port Browning to the north shore of Shingle Bay, from where it is believed to continue up Ganges Harbour to connect with the *Ganges Fault* on Saltspring Island. The *Ganges Fault* trends northwesterly close to Booth Inlet, from where it probably extends as far as Ladysmith Harbour.

What is Happening Today?

The structures we have been talking about are thought to have been caused by the collision of *Wrangellia* with the edge of North America about 100 million years ago and by compression of the crust as a result of the accretion of the *Pacific Rim* and *Crescent terranes* against and beneath the island, between fifty-four and forty-two million years ago. These latter events caused the island's rocks to be further squeezed against the Coast Mountains, thus leading to the northwesterly aligned folds of the Gulf Islands and the development of the Cowichan Lake, Beaufort Range and other similarly oriented fault systems. But what is happening today?

As we discussed earlier, the floor of the Pacific Ocean lying to the east of the Juan de Fuca Ridge (page 25) is converging toward and being subducted beneath the western edge of the continent. Thus, the same processes that led to the accretion of the *Pacific Rim* and *Crescent terranes* are active today and have similar effects. As the oceanic crust spreads away from the Juan de Fuca Ridge, it becomes covered by a blanket of sediments formed from the constant rain of fine muds settling out of the ocean. By the time the plate has spread as far as the westward-moving continent, this blanket of sediment has become about 3 km thick. As the two plates converge upon one another, the blanket of sediment is scraped off the subducting oceanic plate and becomes accreted to the edge of the continent. In the process, the once horizontal layers of sediment become folded and faulted in much the same manner as the strata of the Gulf Islands.

HISTORY OF MINING ON VANCOUVER ISLAND

At the time of writing (1993), only two metal mines are operating on Vancouver Island. These are Island Copper, producing copper and molybdenum from an open pit near Port Hardy, and Westmin Mines, which produces lead and zinc together with a multitude of minor metals from underground operations at the south end of Buttle Lake. Coal, which was the mainstay of the mining industry on the island for a hundred years, is currently being extracted only from the Quinsam coal field southwest of Campbell River. Though there has been active exploration for gold, there is no significant commercial production on Vancouver Island at present. Limestone is mined at Benson Lake for processing into a variety of industrial fillers, and marble is being extracted at Bonanza Lake.

COAL

Coal was first mined from the *Nanaimo Group* by the Hudson's Bay Company in 1849 on northern Vancouver Island, where, over about three years, some ten thousand tonnes were produced to supply the post of Fort Rupert near present-day Port Hardy. This operation was abandoned when richer deposits were developed within the *Nanaimo Group* near Nanaimo in 1854. The principal coal fields of the Nanaimo area were at Wellington, a few miles north of Nanaimo, and at Extension and South Wellington, a short distance to the south. The mines in these fields operated until the thirties, when declining markets of the depression years and exhaustion of the best coal seams forced their closure; during World War II some subsidized coal mining took place in the Nanaimo area. The other major coal field, also in *Nanaimo Group* strata, was in the Cumberland area, where mining began in 1870 and continued until the last mine at Tsable River closed in 1960. Throughout the entire period the coal mines of the *Nanaimo Group* yielded some seventy-two million tonnes of bituminous coal.

The terraced walls of the open-pit Island Copper mine near Port Hardy bear testament to the orderly and efficient methods necessary in these kinds of mining operations.

COPPER

The first metal mining on Vancouver Island was for copper. In 1863, Captain Jeremiah Nagle sank a shaft and recovered a small amount of copper ore in a sea cliff of *Metchosin Igneous Complex* rocks, near Iron Mine Bay, in what is now East Sooke Park (Locality 13). In 1897 claims were staked on copper mineralization in Paleozoic rocks of the *Sicker Group* on the west slope of Mt. Sicker, north of Duncan. By 1900 the Leonora mine was in production and a townsite was laid out. By 1901 a narrow gauge railway had been completed to connect with the E & N Railway, and ore was shipped to Ladysmith for transhipment to a smelter at Tacoma, Washington. Subsequently, the railway was extended to Crofton, where a new smelter had been built. By 1903 two other mines, the Tyee and the Richard the Third, also were shipping copper ore from Mt. Sicker. The mines closed in 1908, though the smelter at Crofton continued to operate on ore shipped from the Britannia mine at the head of Howe Sound on the mainland.

In recent years major production has come from the Island Copper mine near Port Hardy, where a large, low-grade copper/molybdenum deposit is being mined from an open pit, excavated in Jurassic igneous rocks of the *Bonanza Group*. The mine opened in 1971 with stated reserves of 257 million tonnes of 0.52 percent copper and 0.17 percent molybdenum, as well as significant amounts of precious metals. At the south end of Buttle Lake, the Lynx-Myra-Price and HW ore bodies produce copper, zinc, lead and precious metals from Paleozoic rocks of the *Sicker Group*. In 1983 combined total reserves were estimated at twenty-one million tonnes of 1.5 to 2.2 percent copper, 5.3 to 7.6 percent zinc, 0.3 to 1.1 percent lead and 37.7 to 109 grams per tonne silver and 2.0 to 2.4 grams per tonne gold. Over the years several other small mines have produced copper ore, usually with some gold; these include the Sunro mine at River Jordan, the Mt. Washington open-pit mine, Blue Grouse mine near Cowichan Lake, and King Solomon mine in the Koksilah River valley.

GOLD

The early history of the city of Victoria is intimately associated with the gold rush to the Fraser River in 1858 and to the Cariboo fields in 1863. **Placer** gold was discovered near Victoria on the Goldstream River in 1858 and significant deposits were found at the mouth of the Leech River in 1864, the latter leading to a short-lived gold rush. Subsequently, placer miners worked China Creek near Alberni, the Oyster River near Courtenay and streams in the Zeballos area, and even recovered fourteen hundred ounces of gold from black beachsands at Wreck Bay in what is now Pacific Rim National Park. On southern Vancouver Island, stream placer deposits have been worked in Loss Creek, Leech River, Sombrio River, River Jordan, San Juan River, Muir Creek, Sooke River and other coastal streams. During the late nineteenth century, yields of placer gold from workings in the Leech, Jordan and San Juan rivers were estimated to have been in excess of $200,000. Sombrio Point, the site of an extensive hydraulic placer mining operation in 1910, recently has been reexamined.

Amounts of gold recovered on Vancouver Island are minor compared with the rich placer deposits of the mainland. In the Zeballos area on the northeast coast of the island, prospectors following up the placer workings discovered gold mineralization in bedrock; several deposits of **lode** gold produced ore in the 1920s and 1930s. Half of the production came from the Privateer vein at the White Star property, and, up to 1948, 287,800 ounces of gold and 124,700 ounces of silver were extracted from 651,000 tonnes of ore. On southern Vancouver Island, lode-gold deposits are associated with quartz veins in the *Leech River Complex* of the *Pacific Rim Terrane* and within the *Sicker Group* of *Wrangellia*. Extensive exploration, including drilling, has as yet failed to identify economically viable deposits. Undoubtedly it is these gold-quartz veins that have been the source of gold placers in many of the streams draining the southernmost part of the island.

IRON/COPPER

Several magnetite deposits have been mined as a source of iron ore. These usually contain some copper in the form of chalcopyrite. Former producing mines are located south of the Tofino-to-Ucluelet highway on the Kennedy River, on Iron River near Quinsam Lake west of Campbell River and on Benson Lake near Port McNeill. These deposits were of limited extent and the products were shipped to steel mills in Japan.

LIMESTONE

Limestone quarries for the production of quicklime and cement were active on the Saanich Peninsula, at Bamberton below the Malahat highway and near Cobble Hill. The Saanich quarry has been reclaimed and is well known to tourists as the Butchart Gardens.

SAND AND GRAVEL

Deposits of sand and gravel laid down by glacial meltwater are abundant on Vancouver Island, and many have been developed for concrete, road building and other construction needs. One of the largest gravel pit operations in British Columbia is the Metchosin Pit in Colwood (Locality 19).

BUILDING STONE

From the turn of the century until the 1930s, British Columbia produced a wide variety of building stones, both for use in the province and for export. On Vancouver Island, marble was quarried from the Quatsino Form*ation* in the Nootka Sound region. Similar marble is currently being exploited near Bonanza Lake. Perhaps the most famous quarry is the one located on Haddington Island, east of Port McNeill. This fine-grained, gray volcanic rock has been used in many historic buildings in Victoria, including those of the Provincial Legislature. Sandstone was extracted from numerous quarries on Denman and other islands of the Gulf Islands up to the 1920s and can be seen in many buildings in Victoria (eg. Carnegie Building) and Nanaimo (post office), as well as farther afield (eg. San Francisco Mint). An excellent discussion of building stone in Victoria has recently been published by Z.D. Hora and L.B. Miller (see *Additional Reading*).

EARTHQUAKES ON VANCOUVER ISLAND

The same kinds of forces that have shaped Vancouver Island throughout the past 375 million years are active today. The formation of mountains, faults, folds and the intrusion of molten magma into the crust all result from forces acting upon and within the earth's crust. That such forces are still active is demonstrated by the frequent occurrence of earthquakes.

Vancouver Island is situated in the area of highest earthquake risk in Canada. More than two hundred earthquakes are recorded each year on the lower mainland and the island. Those who live in the Greater Victoria area are well aware of earthquakes, and many have felt their effects. Dishes rattle. Suspended objects sway. Sometimes there is even a gunshot-like noise. In these instances the earth is responding to an accumulation of stress, due mainly to the fact that we live in a region where two of the earth's tectonic plates are interacting with each other. As discussed on page 25, new sea floor is being created along the *Juan de Fuca Ridge* from where the oceanic crust of the *Juan de Fuca Plate* spreads eastward towards Vancouver Island. Vancouver Island, being part of the *North American Plate*, is moving westward as a consequence of sea-floor spreading in the Atlantic Ocean. Where the two approaching plates meet, the oceanic crust of the *Juan de Fuca Plate* descends, or is subducted, beneath the *North American Plate*. It is the creation of oceanic crust along the *Juan de Fuca Ridge* and its subduction beneath Vancouver Island that is responsible for much of our seismic activity.

The **magnitudes** of earthquakes that occur on Vancouver Island are generally low, in the order of two to four on the **Richter scale**. The Richter scale is a measure of the magnitude of an earthquake as related to the amount, or amplitude, of vertical ground motion. The scale is logarithmic, meaning that the recorded amplitude of a magnitude seven earthquake is ten times greater than one of magnitude six, a hundred times greater than one of magnitude five, and so on. The amplitude of vertical ground motion increases with increasing magnitude. For example, a magnitude three tremor, common in the Greater Victoria area, would result in a displacement of the ground in the order

The Modified Mercalli Scale

I Not felt except by a very few under especially favourable circumstances.

II Felt only by a few persons at rest, especially on upper floors of buildings. Delicately suspended objects may swing.

III Felt quite noticeably indoors, especially on upper floors of buildings, but many people do not recognize it as an earthquake. Standing motor cars many rock slightly. Vibration like passing truck. Duration estimated.

IV During the day felt indoors by many, outdoors by few. At night some awakened. Dishes, windows, doors disturbed; walls make creaking sound. Sensation like heavy truck striking building. Standing motor cars rocked noticeably.

V Felt by nearly everyone; many awakened. Some dishes, windows, etc., broken; a few instances of cracked plaster; unstable objects overturned. Disturbances of trees, poles, and other tall objects sometimes noticed. Pendulum clocks may stop.

VI Felt by all; many frightened and run outdoors. Some heavy furniture moved; a few instances of fallen plaster or damaged chimneys. Damage slight.

VII Everyone runs outdoors. Damage negligible in buildings of good design and construction; slight to moderate in well-built ordinary structures; considerable in poorly built or badly designed structures; some chimneys broken. Noticed by persons driving motor cars.

VIII Damage slight in specially designed structures; considerable in ordinary substantial buildings, with partial collapse; great in poorly built structures. Panel walls thrown out of frame structures. Fall of chimneys, factory stacks, columns, monuments, walls. Heavy furniture overturned. Sand and mud ejected in small amounts. Changes in well water. Disturbs persons driving motor cars.

IX Damage considerable in specially designed structures; well-designed frame structures thrown out of plumb; great in substantial buildings, with partial collapse. Buildings shifted off foundations. Ground cracked conspicuously. Underground pipes broken.

X Some well-built, wooden structures destroyed; most masonry and frame structures destroyed with foundations; ground badly cracked. Rails bent. Landslides considerable from river banks and steep slopes. Shifted sand and mud. Water splashed over banks.

XI Few, if any, (masonry) structures remain standing. Bridges destroyed. Broad fissures in ground. Underground pipelines completely out of service. Earth slumps and land slips in soft ground. Rails bent greatly.

XII Damage total. Waves seen on ground surfaces. Lines of sight and level distorted. Objects thrown upward into the air.

of a millimetre, a five in the order of centimetres, and a seven about 2 m. A magnitude nine earthquake would produce a vertical displacement of the ground of tens of metres.

Another measure of earthquake effect is **intensity**. The intensity of an earthquake at any location depends upon the magnitude of the earthquake, the distance of the location away from the epicentre and the nature of the ground underlying the site. The intensity scale, called the *Modified Mercalli Scale*, is a measure of the degree to which an earthquake affects people,

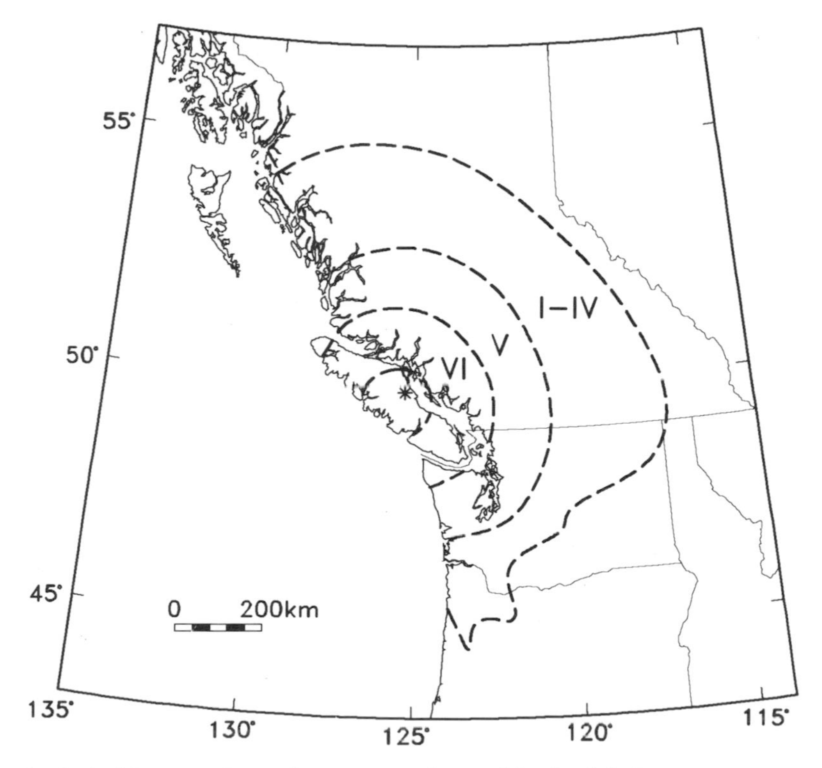

The dashed lines are lines of approximately equal levels of shaking intensity resulting from the 1946 Vancouver Island earthquake. Each line separates regions (I to VII) which were affected according to the Modified Mercalli Scale of earthquake intensity.

property and the ground surrounding a given site. The scale is shown on page 54.

Shortly following a significant earthquake in British Columbia, questionnaires are sent to people in the affected area. The answers to these questions enable seismologists to estimate the intensity of the quake and to construct **isoseismal maps** such as the one illustrated above. The contours, centred upon the epicentre, define the distribution of the intensity zones of the *Modified Mercalli Scale*. Although individual replies to the questionnaire can be highly subjective and variable even within the same zone, when taken all together and compiled into a map, the replies provide seismologists with useful information which can be applied to regional earthquake studies in a wide variety of ways.

The amount of energy released by an earthquake also can be estimated from a knowledge of its location and magnitude. With each unit increase in magnitude, the amount of energy released

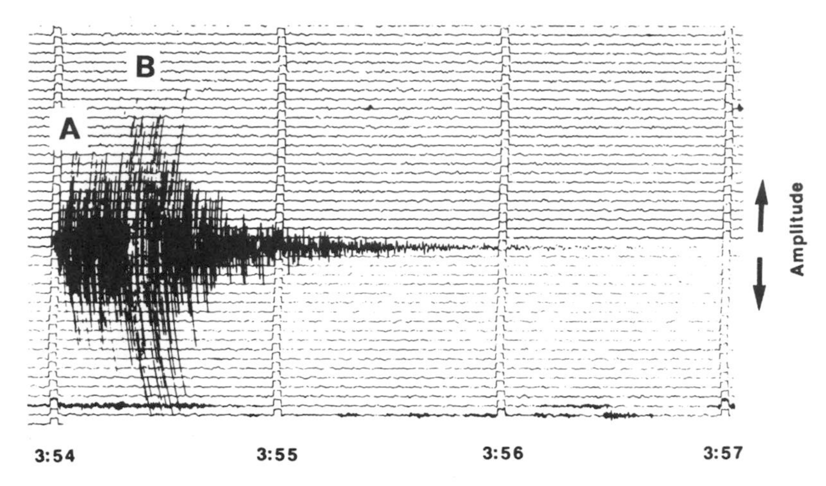

This seismogram was recorded by a seismograph located at the Gonzales Observatory at 3:45 pm PST, on October 24, 1989, and telemetred to the Pacific Geoscience Centre. Based upon this and recordings of the same event at other seismic stations, it was determined that the epicentre of the earthquake was 25 kilometres east of Ucluelet and 45 kilometres southwest of Port Alberni (latitude 48.94°N; longitude 125.16°W). The magnitude of the earthquake was calculated at 3.9 on the Rich]ter Scale and occurred at a depth of 35 kilometres below the surface, within the subducting Juan de Fuca Plate.

increases by a factor of thirty-two (not by a factor of ten as is often erroneously stated in the press). For example, a magnitude five earthquake releases about thirty times more energy than one of magnitude four and some one thousand times more than a magnitude three quake.

The above llustration shows the record of a typical earthquake of magnitude 3.9, which occurred at a focal depth of about 35 km below Barkley Sound on the west coast of Vancouver Island at 3:54 pm, on October 24, 1989. The earthquake lasted for about two minutes. Notice the difference in amplitude between the **primary**, or compressional wave, at A, and the **secondary**, or shear wave, at B. The difference in arrival time at the recording station between these two different types of waves provides a measure of the distance of the epicentre from the recording station, in this case the Gonzales Observatory in Victoria (Locality 8). Three or more recording stations are required to fix an epicentre precisely. The calculated depth to the source of this earthquake indicates that it occurred within the upper part of the subducting oceanic crust of the Juan de Fuca Plate.

The Geological Survey of Canada's Pacific Geoscience Centre (PGC), located on the site of the Institute of Ocean Sciences at

Accessible only by helicopter, a seismic recording station is situated atop Mt. Septimus in Strathcona Provincial Park, from where digitally recorded earthquate data are transmitted to the Pacific Geoscience Centre at Patricia Bay.

Patricia Bay near Sidney, is the hub of a network of sixty seismological stations monitoring earthquake activity throughout western Canada. Data from this network of remote, digital recording seismographs are telemetred to PGC, providing information on the magnitude, epicentre location and depth of earthquakes. A network of forty strong-motion seismographs has been deployed to record precisely earthquake motions great enough to cause damage. When possible, following an earthquake of significant magnitude, seismologists at PGC use portable seismographs to monitor aftershocks in epicentral regions. These data are used to provide information on the nature of the stresses that caused the earthquake.

PART TWO

FIELD GUIDE TO THE GEOLOGY OF THE GREATER VICTORIA REGION

INTRODUCTION

The city of Victoria and its surrounding communities owe their locations to some 350 million years of geological history. Whereas most of the city is built upon rocks that once lay about 20 km below the surface of Vancouver Island, the communities of Metchosin and Sooke owe their place to a piece of sea floor that was rammed underneath Vancouver Island about forty-two million years ago. People living on Mt. Newton have homes built upon frozen magma that was intruded into Vancouver Island crust about 170 million years ago. A walk along the shores of Gonzales Bay or a drive up the Malahat from Goldstream Park carries you across ancient deep-ocean sediments, which also were emplaced beneath the island along faults. While beachcombing along the northern Saanich Peninsula or in the Gulf Islands, you can see sedimentary strata that have been bent into folds or broken by faults. At Nanaimo these same strata contain coal, which for many years sustained the economy of this region. Or, while simply wandering about Victoria, you will notice that most bedrock outcrops are grooved or scratched in a general north-south direction, testaments to the fact that, up until thirteen thousand years ago, the city and all of Vancouver Island were covered by up to two kilometres of glacial ice.

Throughout most of the Capital Region, particularly where traffic has been active for many years, bedrock outcrops are covered with dark soot, which renders them difficult to examine. The best places to study the rocks are along the rocky coast where they are continually washed by the tides, or upon prominent hilltops, away from the grime and the dust.

The oldest rocks in the region occur within a ten-kilometre-wide zone trending southeasterly from the San Juan – Koksilah River valleys, across Finlayson Arm and through much of Greater Victoria to the sea coast between Uplands and Fort Rodd Hill parks. These are metamorphic rocks – originally igneous and sedimentary in origin, but which late in their histories were subjected to high heat and pressure such that their constituent minerals were changed in chemical composition and crystal structure. There are two kinds. One has been named the *Wark Gneiss* and mainly consists of dark-coloured, finely banded, or

foliated, rock composed of the minerals **hornblende** and **plagioclase** feldspar. At many localities no banding is evident and the rock is called an **amphibolite**, a rock in which the mineral hornblende is most abundant. Good exposures of the *Wark Gneiss* occur on Mt. Tolmie (Locality 1), along the Trans-Canada Highway west of the city, as coastal outcrops along the shores of Portage Inlet and Esquimalt Harbour and at the intersection of Menzies and Belleville streets at Confederation Fountain. At this last locality an intricate network of quartz veins has been injected into the gneiss, which has been prominently grooved by glacial ice. The other type of metamorphic rock in the Greater Victoria area is called the *Colquitz Gneiss*, consisting of foliated light- and dark-banded rocks composed of the minerals **biotite**, hornblende, quartz and plagioclase. At many localities throughout the city these two gneisses are mixed together such that one encloses or is interlayered with the other. Excellent exposures of the *Colquitz Gneiss* occur at Cattle Point (Locality 3), where the layering has been tilted vertically so that you can readily see that these were originally sedimentary strata before they were metamorphosed into gneiss. At widely scattered localities throughout the region underlain by these metamorphic rocks, small bodies of **marble** (metamorphosed limestone) are enclosed by the gneisses. Examples occur at Bamberton, where for many years cement was produced from the limestone, at the Butchart Gardens and along Highway 1A close to the head of Esquimalt Harbour. Along the shore of Cordova Bay (Locality 5) are younger limestones interstratified with submarine lavas thought to belong to the Upper Triassic *Karmutsen Formation*, one of the most widespread and thickest formations of Vancouver Island. Elsewhere, such as at Coles Bay (Locality 6), Ten Mile Point (Locality 7) and throughout Mt. Newton and Bear Hill, Early Jurassic granitic rocks form "salt and pepper" exposures of molten magma that was injected into Vancouver Island about 170 million years ago.

All of these aformentioned rocks belong to the terrane called *Wrangellia*, the large piece of crust making up much of Vancouver Island, the Queen Charlotte Islands and parts of southeastern Alaska. As discussed in Part One, *Wrangellia* is thought to have become part of North America by about 100 million years ago (pages 16 to 29), following extensive travels throughout the ancient Pacific Ocean by the processes of sea-floor spreading and plate tectonics.

As a consequence of its accretion to North America, the rocks of *Wrangellia* were compressed and uplifted to form a range of mountains extending throughout much of Vancouver Island. The erosion of this range and other mountain systems flanking the *Georgia Basin* resulted in the accumulation of a thick succession of sedimentary strata, including coal, on top of *Wrangellia*. These rocks form the Cretaceous *Nanaimo Group*, which underlies the east coast of the island and which forms most of the Gulf Islands (Locality 20) and the northern coast of the Saanich Peninsula (Locality 10).

Two other terranes also contribute to the geological architecture of Greater Victoria. One is called the *Pacific Rim Terrane*, which was accreted to *Wrangellia* prior to about fifty-five million years ago and which, in this area, is made up of metamorphic rocks of the *Leech River Complex*, as seen at McNeill and Gonzales bays (Locality 8), at Goldstream Park and along the Malahat Drive (Locality 9) and at Botanical Beach near Port Renfrew (Locality 17). The other piece of crust, called the *Crescent Terrane*, consists of ancient sea floor, formed about fifty-four million years ago and emplaced beneath the *Pacific Rim Terrane* about forty-two million years ago. The *Crescent Terrane* on Vancouver Island consists of lavas and **intrusive rocks** of the *Metchosin Igneous Complex* and is well exposed at Witty's Lagoon Park (Locality 11) and at Aylard Farm and Iron Mine Bay, each being localities in East Sooke Park (Localities 12 & 13). Following accretion of these two terranes to *Wrangellia* (Vancouver Island), a succession of sedimentary strata was deposited across the faults which separated them from each other and from *Wrangellia*. These rocks, called the *Sooke Formation*, are seen at East Sooke Park (Locality 12) and at French, China and Botanical beaches (Localities 14, 15 & 17).

The effects of glaciation of southern Vancouver Island are well displayed at many localities throughout the Greater Victoria area. At Finlayson Point (Locality 4), Cowichan Head (Locality 18) and throughout the Metchosin, Colwood and Langford areas (Locality 19), glacial erosional and depositional features attest to a history of some fifty thousand years when, for much of that time, southern Vancouver Island and the straits of Georgia and Juan de Fuca were entombed in ice.

The geological map shown on page 63 is greatly simplified and based upon one prepared by J.E. Muller for the Geological Survey of Canada. The map depicts the distribution of the different types of bedrock in the Greater Victoria area, both where they are widely exposed and where they are inferred to exist beneath a thick cover of glacial debris (cross hatching). Also shown are the surface traces of the main faults separating *Wrangellia*, the *Pacific Rim* and *Crescent terranes*. North of the

LEGEND FOR THE GEOLOGICAL MAP

QUATERNARY

Unconsolidated glacial sediments overlying bedrock.

TERTIARY

A. **Sooke Formation:** Sandstone, conglomerate.

B. Quartz-feldspar granitic intrusive rocks.

C. **Metchosin Igneous Complex:** Marine and nonmarine lavas, gabbro. = **Crescent Terrane.**

CRETACEOUS

D. **Nanaimo Group:** Sandstone, conglomerate, shale.

CRETACEOUS AND JURASSIC

E. **Leech River Complex:** Metamorphosed sedimentary and volcanic rocks. = **Pacific Rim Terrane.**

JURASSIC

F. **Bonanza Group:** Volcanic rocks.

G. **Island Intrusions:** Granodiorite intrusive rocks.

TRIASSIC

H. **Vancouver Group: Karmutsen Formation:** basaltic lava. **Quatsino Formation:** limestone. Volcanic rocks of possible Paleozoic age.

PERMIAN AND CARBONIFEROUS

I. **Buttle Lake Group:** Limestone, chert, argillite.

DEVONIAN

J. **Wark and Colquitz Gneiss Complexes:** Metamorphosed sedimentary and igneous rocks, possibly originally belonging to the Sicker Group and metamorphosed during Early Jurassic time.

K. **Saltspring Intrusions:** Metamorphosed granitic rocks.

L. **Sicker Group:** Argillite, chert.

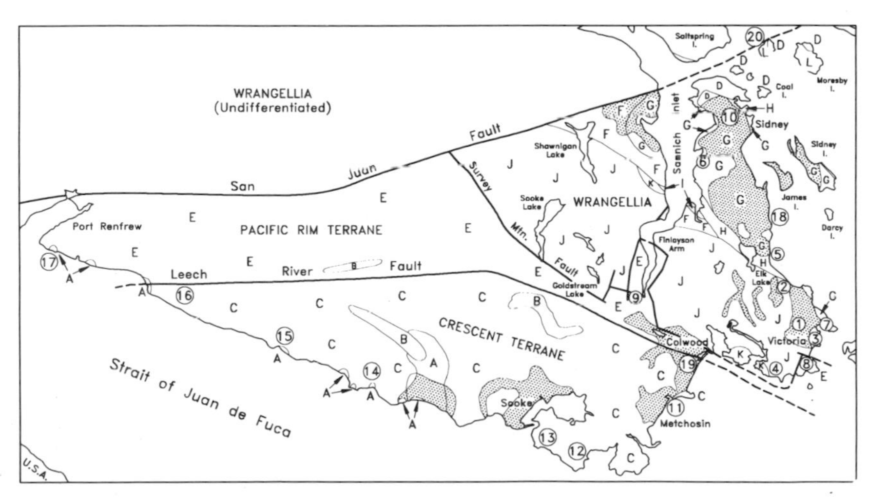

Geological map of southern Vancouver Island, based upon Geological Survey of Canada Map No. 1553A by J.E. Muller (see page 3). Readers are invited to colour this map and its associated legend using the following suggested colour scheme: A = yellow; B = pink; C = orange; D = lime green; E = dark green; F = brown; G = red; H = olive green; I = gray; J = blue; K = purple. The circled numbers are localities described in the text.

San Juan Fault, the several rock formations of *Wrangellia* are not shown, as they are in the triangular region bordered by the Survey Mountain and San Juan faults. Likewise, subdivision of the *Pacific Rim* and *Crescent terranes* into their several component rock types was not considered useful for the purpose of this guide.

The following localities are representative of much of Vancouver Island's geological history. They are arranged in order of decreasing age of the bedrock of which they are formed; some are related and thus are grouped together. The last locality is a tour by B.C. Ferries through the Gulf Islands.

List of Localities:

1. Mt. Tolmie
2. Mt. Douglas
3. Cattle Point
4. Finlayson Point and the Beacon Hill Park Sea Cliffs
5. Cordova Bay
6. Coles Bay
7. Ten Mile Point
8. McNeill Bay, Gonzales Bay and Gonzales Observatory
9. Goldstream Park, Mt. Finlayson and the Malahat
10. Armstrong Point/Allbay Road
11. Witty's Lagoon Park
12. East Sooke Park: Aylard Farm
13. East Sooke Park: Iron Mine Bay
14. French Beach Provincial Park
15. China Beach Provincial Park
16. Loss Creek Park Provincial Park
17. Botanical Beach Provincial Park
18. Island View Beach and Cowichan Head
19. Colwood Delta
20. The Gulf Islands

1.
MT. TOLMIE

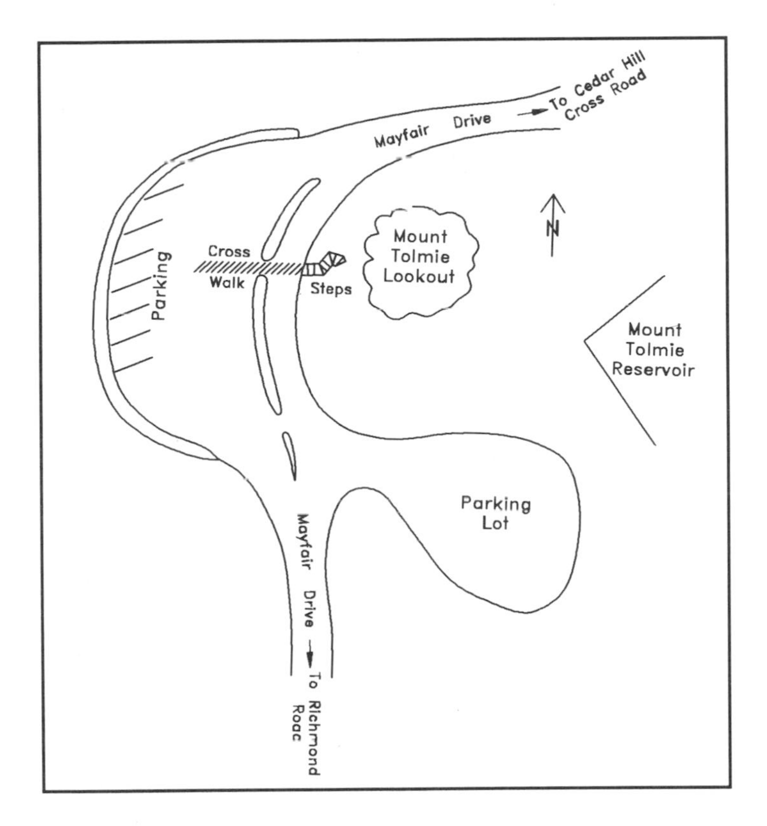

Mt. Tolmie lookout provides a 360-degree view of the southern tip of Vancouver Island and is an excellent point from which to get a general understanding of the principal topographic features of the Victoria area.

Mt. Tolmie viewpoint is reached by following Mayfair Drive from its intersection with Richmond Road or from its intersection with Cedar Hill Cross Road. The map shows the parking areas and other features of the lookout. The viewpoint can be reached also by a pleasant walk up through Mt. Tolmie Park from Cedar Hill Cross Road, following Mayfair Drive and the nearby paths.

From the top of Mt. Tolmie, the conspicuous rocky hill to the north is Mt. Douglas. To the right (east) of Mt. Douglas you can see the cliff on the south end of James Island, which consists principally of gravels of the *Quadra Formation* (see Locality 18). Farther to the right you can see other sand and gravel cliffs on the west shore of Sidney Island. These sands and gravels probably formed a continuous blanket of sediments and included those exposed in the cliffs at Cowichan Head and Island View Beach, hidden behind Mt. Douglas but clearly visible from the Mt. Douglas viewpoint (Locality 2). On a clear day the Coast Mountains on the mainland can be seen on the skyline in line with James Island. These mountains north of Vancouver and in the vicinity of Howe Sound were the source of glaciers that flowed down the Strait of Georgia and over the site of Victoria. Some of the granitic erratics found around the city may have been carried by the glaciers from these mountains.

To the left (west) of Mt. Douglas you can identify Little Saanich Mountain (Observatory Hill), topped with the conspicuous dome of the Dominion Astrophysical Observatory. Again on a clear day, on the skyline between Mt. Douglas and Little Saanich Mountain, you can see the mountains of central Vancouver Island north of the Cowichan Valley. During the onset of *Fraser Glaciation* (see page 33) glaciers formed on these mountains and flowed down the Cowichan Valley and across the north end of the Saanich Peninsula. Meltwater flowing from the front of these glaciers deposited the sand and gravel exposed in the cliffs on James Island and Sidney Island, as well as in gravel pits on the peninsula.

Farther to the left of Little Saanich Mountain, the steep cliffs of the Malahat ridge lie behind Mt. Work, the highest point on the Saanich Peninsula. Looking to the west you can see the rounded shapes of the Sooke Hills, part of the Victoria Highland, which descends from the mountains of central Vancouver Island to the southern tip of the island at Race Rocks. Closer at hand, you can see the waters of Victoria and Esquimalt harbours, and if the lighting is right (preferably in the morning), on the far side of Royal Roads where freighters are sometimes anchored, you will see the large gravel pit and the distinct line marking the surface of the *Colwood Delta* (Locality 18). To the south are the Olympic Mountains on the south side of Juan de Fuca Strait. If you follow the crests of the Olympic Mountains along to the left (east) you will see how they gradually descend to a lowland. This lowland marks the entrance to Puget Sound, and occasionally, when the light is right and the air is particularly clear, you can see Mt. Rainier south of Seattle in line with the junction of

These three views from atop Mt. Tolmie form a part of a panorama. To the north (top) Mt. Douglas (Locality 2) is seen to the right, and to the left is Observatory Hill, capped by the conspicuous dome of the Dominion Astrophysical Observatory. The middle view shows the distant Gowlland Range west of the city, and at the bottom, the snow-capped peaks of the Olympic Mountains rise above the Strait of Juan de Fuca.

Beneath a darkening evening sky stands the snow-encrusted edifice of Mt. Baker, one of several dormant volcanoes extending up the coast from northern California to southern British Columbia.

the lowland and the Olympic Mountains.

Looking to the east you can see San Juan Island, one of several American islands forming a group north of the entrance to Puget Sound. In the distance are the Cascade Mountains, extending from northern Washington into southern British Columbia, where they occur east of the Fraser River near Hope. If you are fortunate with a clear day, you will get a spectacular view of Mt. Baker. If the day is hazy or the mountain is partially covered by cloud, look for it in line with the Mt. Tolmie Reservoir, which is the flat concrete slab below you.

Mt. Baker is one of a chain of several volcanoes which extend up the coast of North America from northern California to north of Vancouver. Others in the group include the notorious Mt. St. Helens, Mt. Rainier, Mt. Shasta, Mt. Hood and, in Canada, Mt. Garibaldi, Meager Mountain and Mt. Cayley. The development of this chain of volcanoes is due to the melting of oceanic crust of the Juan de Fuca Plate as it is consumed, or subducted, beneath the edge of the continent (see page 25).

Mt. Baker is about 3,500 m in elevation and lies about 25 km south of the International Boundary. It was originally named La Montana del Camelo in 1790 by the Spanish explorer Manuel Quimper. Loosely translated, this means "Great White Watcher." The local Nooksack people called it Koma Kulshan, meaning

These two photos show characteristic features of the Wark Gneiss. The light-coloured material consists of veins and dykes of quartz and feldspar within the darker-coloured dioritic gneiss rich in the mineral hornblende.

"white steep mountain." However, those names were ignored when Captain Vancouver named it after his third lieutenant, Joseph Baker.

On clear, sunny days the mountain is a spectacular sight. About fifty-two square kilometres of glaciers mantle its slopes. Just below the summit is the active crater, from which varying amounts of steam and hydrogen sulphide gas have been more or less continuously discharged since the crater formed several tens of thousands of years ago. The volcano is currently dormant; however, between 1820 and 1870 it erupted eight times, with a minor eruption occurring in 1975.

Now that you have made a complete 360-degree examination of the landscape, you should turn your attention to the geology on which you are standing. The rocks of Mt. Tolmie, Mt. Douglas and most of the rocks throughout the Greater Victoria area are metamorphic rocks assigned to the *Wark* and/or *Colquitz gneiss complexes*. These two bodies of crystalline rock appear to be interrelated such that the dark, sombre and massive types of the former are commonly interleaved with the lighter-coloured and banded forms of the latter. The *Wark* and *Colquitz gneisses* con-

The dark, angular outcrops along the Trans-Canada Highway west of the city belong to the Wark Gneiss.

sist of a variety of rocks that were metamorphosed during the early part of the Jurassic Period, about 200 million years ago. The original composition of the *Wark Gneiss* may have been the Devonian granitic and volcanic rocks that form the foundation of the island and which comprise part of the *Sicker Group* (see page 17 and 24). The *Colquitz Gneiss* consists of layered dark- and light-coloured rocks that are possibly metamorphosed sedimentary strata, also belonging to the *Sicker Group.*

Of these two metamorphic complexes, the bedrock at the top of Mt. Tolmie belongs to the *Wark Gneiss*. It has a mottled dark- and light-gray appearance. Looking closely, you might see that the rock has an indistinct foliation, caused by the alignment of thin, platy crystals of the mineral hornblende. The light minerals are **quartz** and feldspar, the proportions and compositions of which define the rock as a **diorite**, a kindred of granite. Thus, in terms of composition and metamorphic texture, the rock is called a dioritic gneiss.

Notice that the surface of the rock is marked by shallow grooves. The grooves were made by pebbles and boulders frozen into the base of glacial ice as it moved across the surface of the rock. The alignment of the grooves and ridges indicates that the glacier flowed southward from the Coast Mountains directly toward the Olympic Mountains. Along the shore of Vancouver Island, west of Race Rocks, however, grooves and striations show the direction of ice movement was westward as the glacier turned in a broad arc and flowed out Juan de Fuca Strait to the Pacific Ocean. An additional lobe of ice extended southward from the Strait of Georgia into Puget Sound on the east side of the Olympic Mountains and ended close to the present city of Tacoma.

To see the composition and texture more clearly on freshly broken rock surfaces, go down the steps and turn left along the edge of the pavement to the parking lot. In this exposure, most of the rocks have been broken along myriad polished fault planes on which there are many fine scratches called **slickensides**. At one place a small fault trending towards the northwest and inclined at about 50° toward the northeast truncates lenses of white quartz. Along several of the paths leading away from the summit are outcrops of the *Colquitz Gneiss*, which is better seen at Cattle Point (Locality 3). The *Wark Gneiss* underlies a northwest-trending zone across the Victoria area (see map on page 63), and numerous exposures can be seen in road cuts throughout the city. A good exposure of the gneiss can be found on the shore at the foot of Bowker Avenue at the south end of Willows Beach.

2.
MT. DOUGLAS

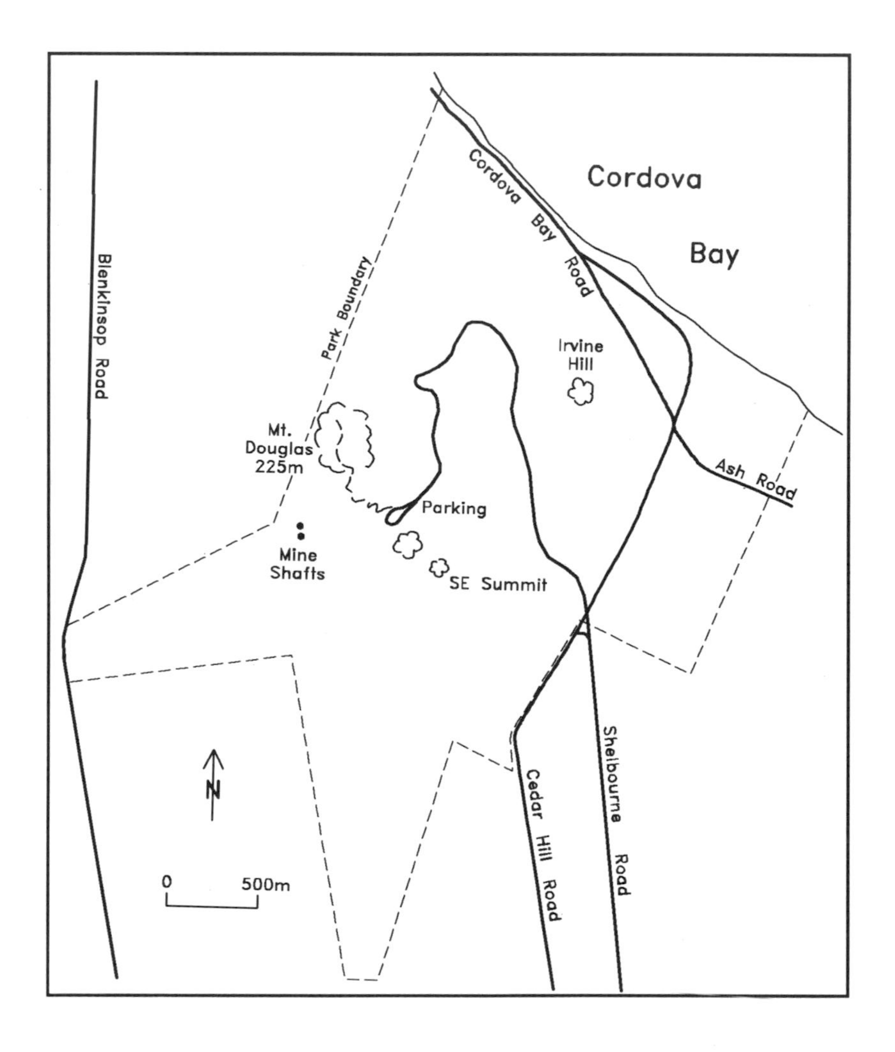

Access to the top of Mt. Douglas is gained via the road entering Mt. Douglas Saanich Municipal Park off Shelbourne Street. In summer this road is sometimes closed during times of severe risk of forest fires. In winter it is often icy. The parking lot is located a short walk below the crest of the mountain.

The rock outcrop on the north side of the parking lot, im-

The effects of differential erosion are shown near the top of Mt. Douglas, where a prominent dyke of quartz and feldspar stands above the more weathered, subdued and lichen-covered metamorphic rocks of the Colquitz Gneiss.

mediately below the trail leading to the top of the mountain, shows evidence of two directions of glaciation. Closest to the tarmac you can see prominent glacial scratches that appear to have sustained a greater degree of weathering than those seen at most other localities throughout the city (eg. Finlayson Point, Locality 4). The direction of these scratches is 195° (or 015°, depending upon your preference). Above the scratches you also can see prominent glacial grooves which have an azimuth of 230° (or 050°). This difference in trend of 35° may have been caused by local differences of flow direction within the same ice sheet, or by two different stages of glaciation. The weathered appearance of the scratches enticingly suggests the latter, meaning that the scratches were formed during an earlier stage of glaciation than that which produced the grooves. Evidence of earlier stages of glaciation is rare on Vancouver Island, although some is provided by the *Muir Point Formation* at Cowichan Head (Locality 18).

Another splendid panoramic view can be obtained from the viewpoint at the top of Mt. Douglas, where a map of the region is engraved upon a bronze plaque atop a concrete column. Around the periphery of the plaque are the viewing directions to many distant features visible throughout an arc of 360°. On especially clear days, and with appropriate atmospheric conditions, the volcanic cone of Mt. Rainier, about 220 km to the southeast, can be seen by sighting between the directions indicated to Mt. Tolmie and Seattle (bearing 155°; for more on these volcanoes see the discussion on Mt. Tolmie, Locality 1). Closer at hand

Two directions of glaciation are displayed by this one outcrop of Colquitz Gneiss at the parking lot near the top of Mt. Douglas. In the lower half of the photo, the nearly horizontal dark and light bands represent scratches caused by ice moving on a bearing of 195° (small ruler). In the upper part of the photo the notebook is aligned parallel to glacial grooves with a trend of 230°. Are these the product of one or two episodes of glaciation?

From the top of Mt. Douglas, the view to the north reveals Cordova Bay, the cliffs of Cowichan Head, and James Island (Locality 18).

and slightly to the left of the direction indicated to Blenkinsop Lake you can see the Metchosin gravel pit close to the shore of Royal Roads (bearing 230°). This gravel pit has been excavated into sediments of the *Colwood Delta*, built from glacial outwash when glaciers melted and retreated from southern Vancouver Island about thirteen thousand years ago (see Locality 19). Also clearly visible are the Dominion Astrophysical Observatory atop Little Saanich Mountain (Observatory Hill) and the Malahat Range beyond. To the north are views of Cowichan Head along the shore of Cordova Bay and nearby James Island (Locality 18).

Mt. Douglas, like Mt. Tolmie, is a hill carved by glacial ice. Unlike Mt. Tolmie, which is sculpted from *Wark Gneiss*, Mt. Douglas is formed from the *Colquitz Gneiss*, a kindred to the Wark but of different original materials prior to their metamorphism. With rare exceptions, the rounded, lumpy outcrops of the *Colquitz Gneiss* of Mt. Douglas are everywhere covered with dark lichens and are thus difficult to observe; discussion of these rocks is left to that of Cattle Point (Locality 3), where they are well exposed. Observable, however, are several light gray to white dykes cutting through the rock. These dykes are composed of feldspar and quartz which were injected into their host rocks during the last stages of metamorphism about 200 million years ago.

3.
CATTLE POINT

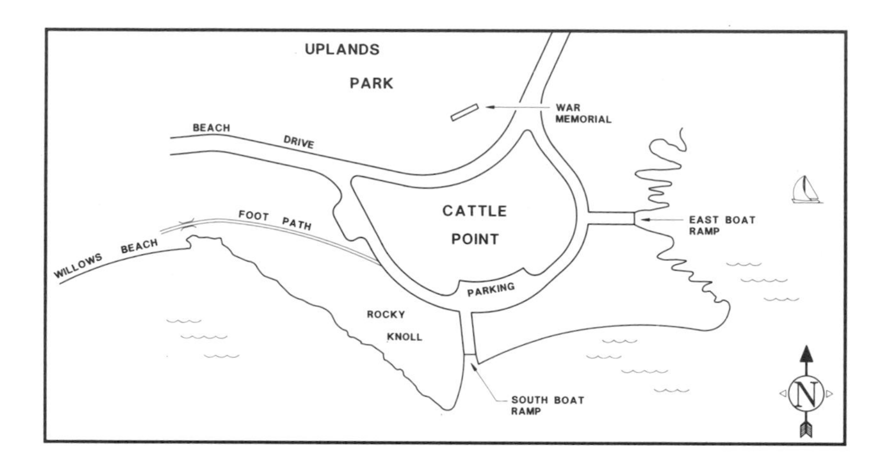

Cattle Point is part of Uplands Park in the municipality of Oak Bay, and can be reached from Beach Drive north of Willows Beach. A convenient place to start is the parking lot near the south boat ramp.

The bedrock consists of metamorphosed sedimentary and volcanic rocks of the *Colquitz Gneiss*, probably originally of Paleozoic age (*Sicker Group*, see page 17) but metamorphosed about two hundred million years ago, during the early part of the Jurassic Period. The gneiss is characterized by alternating bands of light and dark rock layers which commonly are contorted and faulted. Similar rocks are widespread throughout the Victoria area and are intimately intermixed with *Wark Gneiss.*

Standing near the south boat ramp at Cattle Point and looking to the west toward Willows Beach, you see smooth, rounded rock surfaces with traces of grooves, indicating the direction of the glacier ice movement. Walking upon this surface you will notice that, although it was polished smooth by the glaciers, it has since been weathered to a rough texture. Fine, light-coloured, parallel, washboard-like ridges, a few millimetres high, mark the more resistant layers of quartz and plagioclase feldspar, while the softer, more easily weathered, darker minerals, such as hornblende, have been partly removed by frost, wind and rain, thus leaving narrow grooves.

The view to the east from the rocky knoll beside the south boat ramp at Cattle Point shows the character of shoreline and intertidal outcrops of the Colquitz Gneiss.

Walking up to the highest point on the rock knoll north of the boat ramp, you will notice a marked change in the character of the rock surface. Instead of the smooth, rounded surfaces on the main part of the knoll, the south side consists of steep to nearly vertical miniature cliffs, a metre or so high, which form steep slopes extending down to the water. This difference in the character of the rock surface is the result of two different processes of glacial erosion. As glacial ice moved from north to south across the knoll, the north side was eroded by the process of abrasion. Sand, gravel and boulders, embedded in the base of the glacier, acted like coarse sandpaper, smoothing and polishing the rock surface. On the south side of the knoll the rock was eroded by the processes of frost wedging and plucking. Blocks of rock, bounded by randomly oriented fractures, became frozen to the base of the moving ice and were pulled away as the glacier moved on, leaving a rough surface like that found in a rock quarry. These two processes of erosion have combined to produce a characteristic form of glacial landscape known as **roche moutonnée**. Numerous examples of this topography can be seen throughout the Victoria area (Locality 4).

Standing on the high point of the knoll and looking to the northeast, the detailed configuration of the shoreline will vary depending on the state of the tide. If it is a clear day you will see Mt. Baker on the horizon. Along the shoreline are a variety of features characteristic of the metamorphic rocks comprising the *Colquitz Gneiss*. The rocks are of hornblende quartz diorite composition, fine to medium grained and with a steeply inclined foliation and metamorphic layering, generally, but not everywhere, trending northwest-southeast. The surrounding outcrops

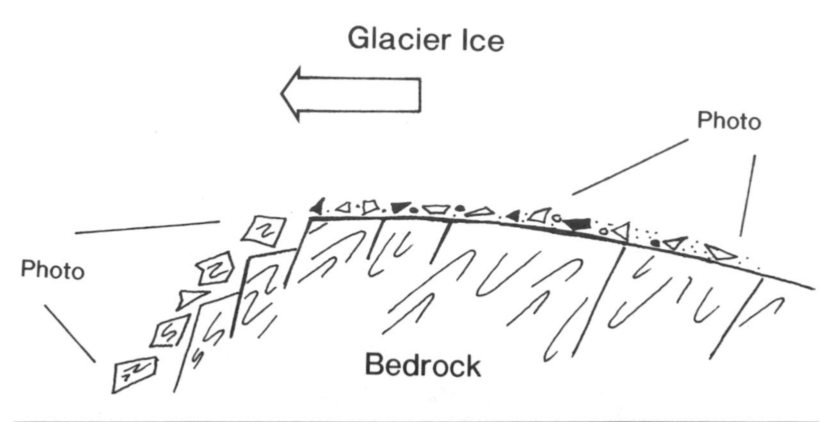

The formation of a roche moutonnée is shown by the drawing and associated photos of the knoll near the south boat ramp at Cattle Point. The smooth, rounded and striated surface (middle) is due to the abrasive action of sand, gravel and boulders embedded in the base of southward-moving (right to left) glacier ice as it crossed the knoll (drawing). On the downstream side of the knoll the rock was eroded by the processes of frost wedging and plucking of the fractured rock, leaving a rough and irregular surface (bottom).

The effects of differential weathering are shown by this "washboard" effect on the surface of the rocky knoll near the south boat ramp at Cattle Point. Alternating vertically inclined layers of light-coloured quartz and feldspar are less easily weathered than those predominantly composed of dark hornblende and micas. The result is that the former stand out in higher relief than the latter.

show complex patterns of faults and folds characteristic of these rocks in the Victoria area. At many locations you will notice how the layers of light-coloured minerals (mostly quartz and plagioclase feldspar) alternate with layers of dark minerals (hornblende), thus forming the gneissic texture of the rock. In some places the layering is cross-cut by wider dykes composed of coarsely crystalline quartz and feldspar.

As you walk along the shore toward the cairn, you will see numerous examples of folded bands and cross-cutting faults. You also will notice boulders as large as 2 m in diameter, composed of light-gray, "salt-and-pepper" granitic rock, which are different from the bedrock upon which they rest. These are erratic boulders of **granodiorite**, transported here by glaciers of the *Fraser Glaciation* (see page 33). They probably came from outcrops of similar rock forming Mt. Newton and Bear Hill on the Saanich Peninsula.

Continuing your walk around the point beyond the cairn, you will see many more examples of the characteristic features of the *Colquitz Gneiss*. Soon you will come into a small bay, which possibly was formed by glacial erosion along a fault that may have been oriented parallel to the direction of the moving ice.

As described in Part One of this guide (page 31), the *Wark* and *Colquitz gneiss complexes*, upon which much of the city is built, were probably once deeply buried beneath the surface, where they formed the foundations to *Wrangellia*. Approximately fifty-four million years ago, rocks of the *Pacific Rim Terrane*, examples of which you see at Goldstream Park, along the Malahat

A quartz/feldspar vein system cuts across vertically inclined metamorphic layering of the Colquitz Gneiss at Cattle Point. The dark layers are composed mainly of hornblende and micas, whereas the thin, wispy light layers are composed of quartz and feldspar. Sudden increases in width of the veins are due to the presence of small fractures which formed after the veins were emplaced in the rocks.

and elsewhere west of Victoria and along the west coast, were rammed beneath the southern and western edges of *Wrangellia*, causing it to be uplifted. The forces of erosion were then able to remove the overlying rocks, thus exposing these once deeply buried metamorphic gneisses at places like Cattle Point.

4. FINLAYSON POINT AND THE BEACON HILL PARK SEA CLIFFS

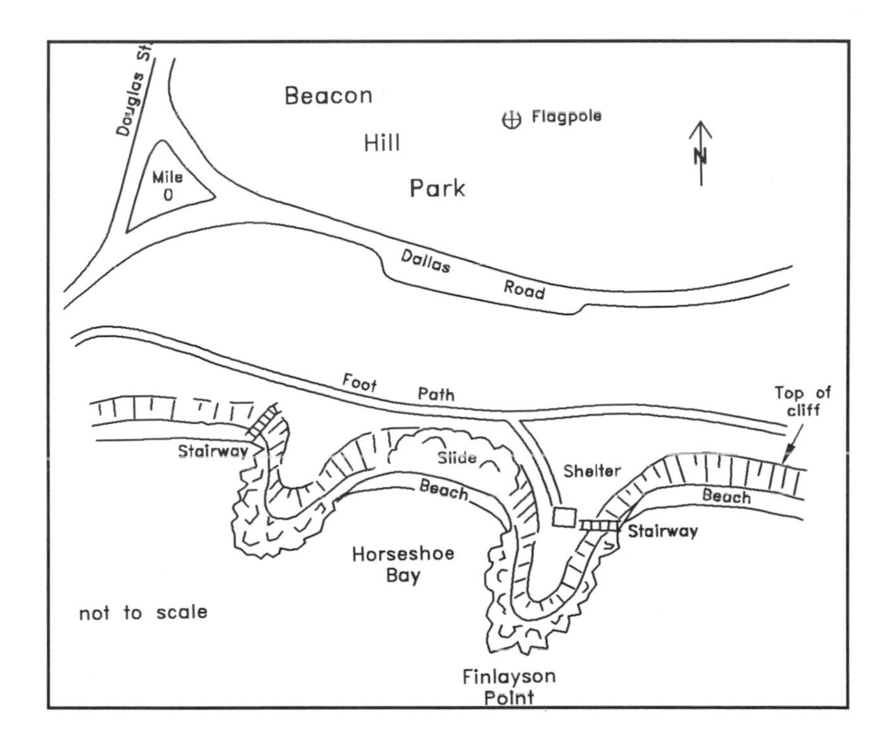

Finlayson Point is located on the shore directly opposite Beacon Hill Park and can be reached from the parking area on Dallas Road just east of its junction with Douglas Street. From the viewpoint above Finlayson Point, look westerly across Royal Roads toward Albert Head and the gravel pit in the *Colwood Delta* (Locality 19). You should easily recognize the level surface of the delta marking the upper limit of marine submergence which occurred thirteen thousand years ago, as the ice of the *Fraser Glaciation* was melting away in this area (see page 33). Closer at hand, the shoreline between Finlayson Point and the breakwater at the entrance to Victoria Harbour consists of rocky points and intervening bays. The rocks have been intensely eroded by glaciers which flowed from north to south across the shoreline. The points consist of resistant rocks,

From atop the sea cliffs beside Dallas Road, the view to the west of the rocky Paleozoic outcrops between Finlayson Point and the entrance to Victoria Harbour is in marked contrast to the distant shore characterized by unconsolidated glacial sediments of the Colwood Delta (Locality 19).

and the intervening bays probably mark zones of faulted and shattered rocks which were more easily eroded by the glaciers. Looking east from Finlayson Point you will see a gently curving beach, backed by steep cliffs that end at rock outcrops on Clover Point. The cliffs consist of unconsolidated sands, gravels and clays which accumulated along the margin of stagnant ice in Juan de Fuca Strait as the *Fraser Glaciation* was waning. Storms coinciding with extreme high tides cause erosion at the base of the cliff, from where material is removed and transported back and forth along the beach. At the foot of the stairs leading down from the boulevard, the beach consists of cobbles of hard igneous rocks which have resisted abrasion by the storm waves.

The bedrock at Finlayson Point consists of fine- to medium-grained, pale greenish-gray and dark gray to black-weathering metamorphosed granitic rocks. The dark-coloured phase is called **gabbro**, a rock containing calcium-feldspar and pyroxene and lesser amounts of other minerals. The light-coloured rocks are granodiorite, consisting of sodium and potassium feldspar, quartz and minor amounts of other minerals. That these two kinds of rock, vastly different in mineral composition, can have formed together and coexist as constituents of a single rock mass has been a subject of controversy among geologists for many years. Regardless of the details, all agree, however, that these kinds of intrusive rocks, called **migmatite**, formed at very deep levels in the crust and are most commonly associated with other metamorphic rocks.

The rocks here are thought to belong to the *Saltspring Intrusions*, which, at other localities where they have intruded the Paleozoic *Sicker Group* (see page 17), have been radiometrically dated at about 360 million years (see explanation on page 95). Here they have been intruded into the *Wark* and *Colquitz gneiss*

The sharply contrasting dark- and light-coloured rock at Finlayson Point is called migmatite, a form of metamorphosed igneous rock that formed at very deep levels in the crust. The dark rocks are gabbro, composed of calcium-feldspar and pyroxene, whereas the light-coloured rocks are granodiorite, mainly consisting of sodium and potassium feldspar and quartz. These are thought to belong to the Saltspring Intrusions of Paleozoic age.

Immediately overlaying the glaciated Paleozoic rocks at Finlayson Point is pebbly and bouldery sand, silt and clay of the Vashon Till, deposited during the Fraser Glaciation, about 17,000 years ago.

complexes, which may be the metamorphosed equivalents of the *Sicker Group*, metamorphism having occurred about 200 million years ago during the early part of the Jurassic Period.

An impressive feature of this locality is the evidence of the erosive power of glaciation. Polished rock surfaces with striations and grooves are conspicuous and good examples of roche moutonnée, and miniature **crag-and-tail** features can be seen. All of these show that the ice moved generally from north to south but, locally, was diverted around small obstructions.

The best preservation of these glacial erosion features is seen close to the base of the landward-retreating cliffs, where the overlying unconsolidated materials have only recently been washed away by storms. Farther seaward from the base of the cliff, the action of waves and weather has removed the polish and striations, but the larger grooves are still easily recognized. Immediately overlying the bedrock are exposures of glacial till consisting of an unsorted mixture of sand, gravel, silt and clay with scattered boulders, which was plastered onto the rock from the bottom of moving ice. It is probable that this is the *Vashon Till*, deposited by the

Spectacular evidence of glaciation is preserved by the rocks at Finlayson Point. Glacial scratches,

grooves,

roche moutonnée

and crag-and-tail are well displayed. Crag and tail, with its steep and rough upstream end and gentle, tapering downstream profile, shows that the ice moved from north (left) to south (right), in the direction indicated by the hammer handle.

glaciers of the *Fraser Glaciation* (see page 36). Higher up the bank is a layer of brown silty clay containing sand and pebbles. This is glacio-marine clay, deposited as the glaciers were melting, when sea level was at the level indicated by the surface of the *Colwood Delta*. In places you may notice a black organic soil containing numerous fragments of broken sea shells and occasional fire-crackled stones. This is probably the remnants of a **midden** from a prehistoric Native encampment on Finlayson Point.

As you continue around Finlayson Point into Horseshoe Bay, you will see the remains of a landslide that occurred in the cliff at the head of the bay in December of 1976. The material that slumped includes Vash*on Till*, glacio-marine clay and the overlying peaty soil. Continuing on around Horseshoe Bay to the rocky point on its west side, you will see more excellent examples of grooves, striations and polished rock surfaces. Bedrock of the *Saltspring Intrusions* continues westward to surround the outer part of the Victoria Harbour and continues on to Duntze Head, forming the east entrance to Esquimalt Harbour.

The foreshore along Dallas Road, forming part of Beacon Hill Park, is characterized by steep cliffs facing upon the Strait of Juan de Fuca. While not immediately apparent, the cliffs are receding by as much as twelve centimetres per year such that, over the past one hundred years, the footpath along the top of the cliffs has been relocated at least three times.

The cliffs expose a variety of glacial sediments, including outwash gravels, tills and materials deposited in contact with ice at or beneath sea level. It is believed that more than twelve thousand years ago, the margin of the Juan de Fuca ice lobe grounded in the near-shore shallow water of the strait, where a series of ice-contact, coalescing sediment fans, or cones, accumulated on the sea floor.

Natural processes associated with ground water, wave action and the nature of the sediments of the cliffs, together with human activity, affect the stability of these bluffs. Ground water seeping from several places at varying rates during the year leads to gullying and small landslides. The action of waves, particularly during winter storms, greatly contributes to erosion and retreat of the bluffs. Westward from Finlayson Point, bedrock helps to retard these effects; however, between Finlayson and Clover points, the lack of exposed protecting bedrock results in significant erosion of the cliffs. The numerous paths crossing the cliff-face concentrate rainwater runoff in channels, leading to accelerated erosion. Since the beginning of the nineteenth century, it is estimated that the cliffs have receeded by about 10 m.

The sea cliffs in Beacon Hill Park along Dallas Road expose ice-contact gravels and till (top) which accumulated adjacent to ice that had grounded in Juan de Fuca Strait about twelve thousand years ago. The view northward along the beach (middle) shows the scalloped form of the cliffs, which are eroding landward at a rate of about twelve centimetres per year. In Horseshoe Bay you can see the effects of a landslide, consisting of Vashon Till and glacio-marine clay that slumped in December of 1976 (bottom).

5.
CORDOVA BAY

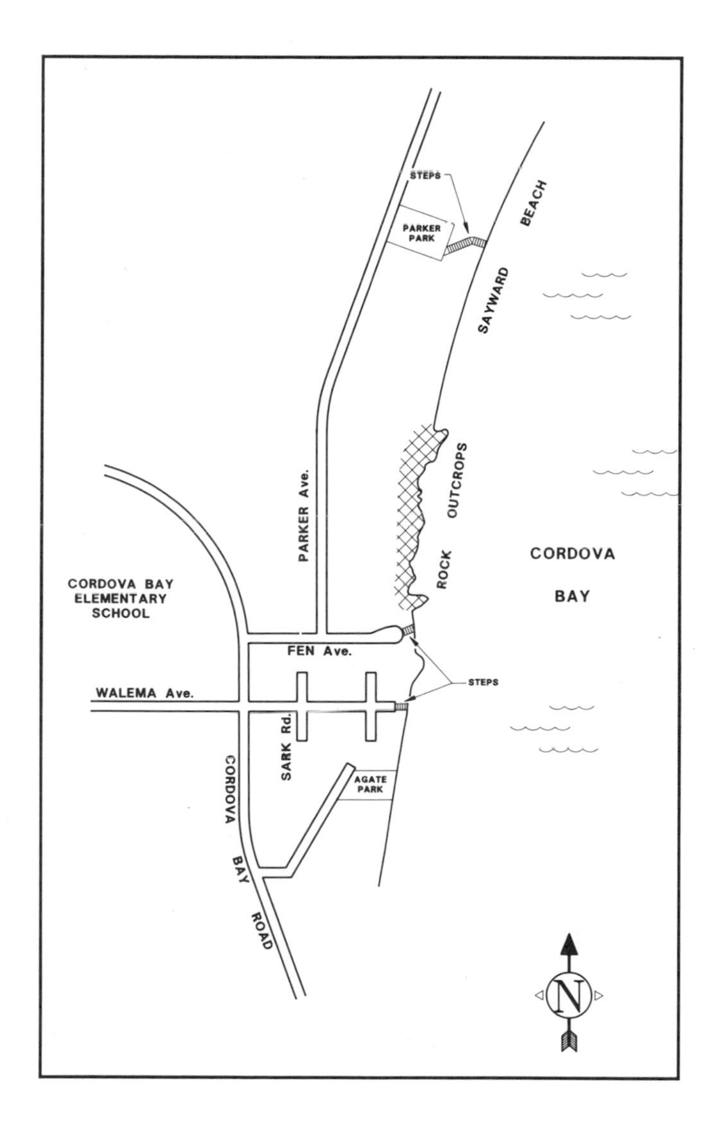

This locality can be reached from Parker Park, as shown on the sketch map. It is inaccessible when the tide is high. From the park, a stairway extends down to Sayward Beach, where the rock outcrops begin a short distance to the south. As you descend, note that the steps and railing show evidence of having been deformed by creep of the underlying soil. All along Parker Avenue the underlying material is soft *Victoria Clay*, which has a tendency to creep and slump during heavy winter rains, or when the toe of the slope is undercut by storm waves. Along the shore you will see how the property owners have attempted to cope with this problem in various ways and with widely differing degrees of success. The rocky shore is composed of dark greenish-gray volcanic rocks, mainly basalt, and lenses of light-gray limestone. The volcanic rocks are mainly extrusive (see glossary, *Igneous Rocks*) lava, consisting of flows and pillow basalt, the latter well displayed about halfway along the outcrop area, where bulbous, pillow-like masses of basalt form a variety of shapes near the shore. Throughout the area of outcrop, these rocks contain curved, swirling veins and lenses of dull-gray calcite and olive-green **epidote**. Farther south these minerals locally define rounded forms up to a metre in diameter, resembling the outlines of pillows. At the south end of the area, smooth glaciated surfaces show cross-sections of pillows, which are outlined with very fine-grained lighter-gray rinds formed from tiny gas-escape pores called **amygdules**; for the most part, these pores are filled with dark-coloured minerals such as epidote and **zeolites**. This crusty rind formed as the result of rapid cooling when the molten lava came in contact with the sea water as it extruded out onto an ancient ocean floor. Much of the basalt has a fragmental structure, resulting from the pillows being broken up as the mass of lava continued to move after it had started to solidify.

Light gray-weathering limestone forms lenses and large pods surrounded by basalt. The limestone is relatively soft and soluble, thus weathered surfaces are smooth, fluted and scalloped. Small **potholes** formed by the abrasion of pebbles rolled by the waves are common above the low-tide line. On freshly broken surfaces the limestone is dark gray, very fine grained and generally massive. Locally, rare crescent-shaped fragments indicate the presence of fossil clams.

The relationship between the basalt and limestone is complex and, at this locality, largely unknown. It is probable, though by no means certain, that the basalt belongs to the *Karmutsen Formation* of Late Triassic age (see page 20). The *Karmutsen*, where it is fully developed throughout central Vancouver Island, commonly contains lenticular masses of limestone in its upper part, similar to the limestone at this locality. On the other hand, the *Karmutsen Formation* generally is overlain by the Upper Triassic *Quatsino Formation*, composed of limestone also very similar in

Bulbous and globular masses of pillow basalt of the Upper Triassic Karmutsen Formation form part of the rocky outcrop along the shore of Cordova Bay.

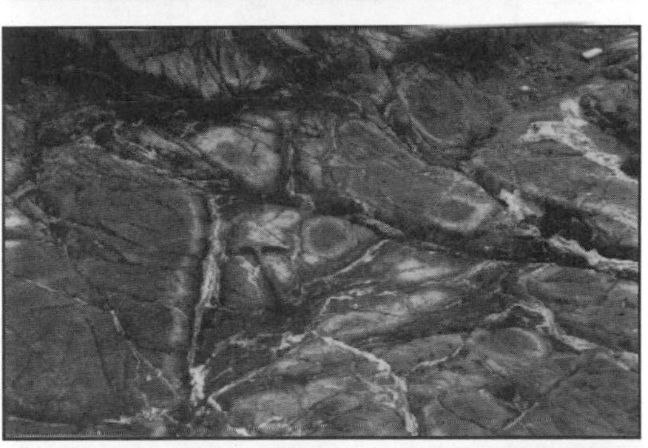

In some places, glaciated and polished surfaces show cross-sections of these pillows, outlined by pale "rinds" formed by gas-escape pores and now filled with minerals such as epidote and zeolites.

Commonly, pillow basalt shatters shortly after extrusion onto the sea floor; the shattered lava then forms pillow breccia, here seen as angular fragments of basalt traversed by numerous fractures filled with epidote and calcite.

appearance to that found within the *Karmutsen Formation* and the limestone exposed here (see page 23). Some of the boundaries between the limestone and basalt are faults, but most are simply irregular depositional boundaries. This irregularity in form and shape of the limestone lenses suggests that the basaltic lavas flowed over a bed of limestone a few tens of metres thick, engulfing parts of it and breaking it into large fragments. This model tends to support the first of the above-mentioned possibilities. The heat of the lava would have tended to cause recrystallization of the limestone and thus the destruction of most of the fossil clamshells and other internal fabrics.

At most places the contact between basalt and limestone is irregular, but locally, steep faults separate the two rock types.

Here, a narrow dyke of dark basalt has been injected into light-coloured limestone.

Throughout the area of outcrop, the basaltic lavas and limestones are cut by basalt dykes lighter in colour than the lavas. They are most clearly seen where they cut through the lighter-coloured limestone. Several trend westerly, perpendicular to the shore, and are steeply inclined. Commonly they are one-half to one metre in thickness and have dark-coloured, fine-grained margins and lighter-coloured central parts. The fine-grained margins are the result of rapid chilling, where the molten dyke-rock came into contact with the cold rock along fractures into which the basalt was intruded. Some dykes are very irregular and merge

Enclosed within Upper Triassic pillow basalt at Cordova Bay are layers of light-gray limestone riddled with potholes (top), which form from dissolution of the limestone, aided by the abrasive action of pebbles (bottom) rolled about by the swirling action waves.

with the basaltic lava, suggesting that the dykes acted as feeders along which the molten lava was erupted onto the ancient Triassic sea floor.

The *Karmutsen Formation* is the thickest and most widespread rock unit on Vancouver Island, and a characteristic formation of *Wrangellia* (see page 20). In addition to its occurrence on the island, it occurs throughout the Queen Charlotte Islands and parts of southeastern Alaska. At Buttle Lake on Vancouver Island, the formation is about 5,500 m thick. Several studies have identified the formation as having formed within an ancient oceanic rift setting, not unlike that of the modern Mid-Atlantic Ridge.

At Gordon Head there are volcanic rocks unlike those at Cordova Bay. Shoreline outcrops in the vicinity of the beach access at the foot of Shore Way consist of intensely broken and moderately metamorphosed black and greenish-gray lavas and minor limestone. These rocks may be of Paleozoic age and kindred to volcanic rocks on San Juan Island, as well as to those underlying Cretaceous sedimentary strata along the northeast shore of the Saanich Peninsula (Locality 10). Similar volcanic rocks occur along the shore of Brentwood Bay and northwest of Elk Lake.

6.
COLES BAY

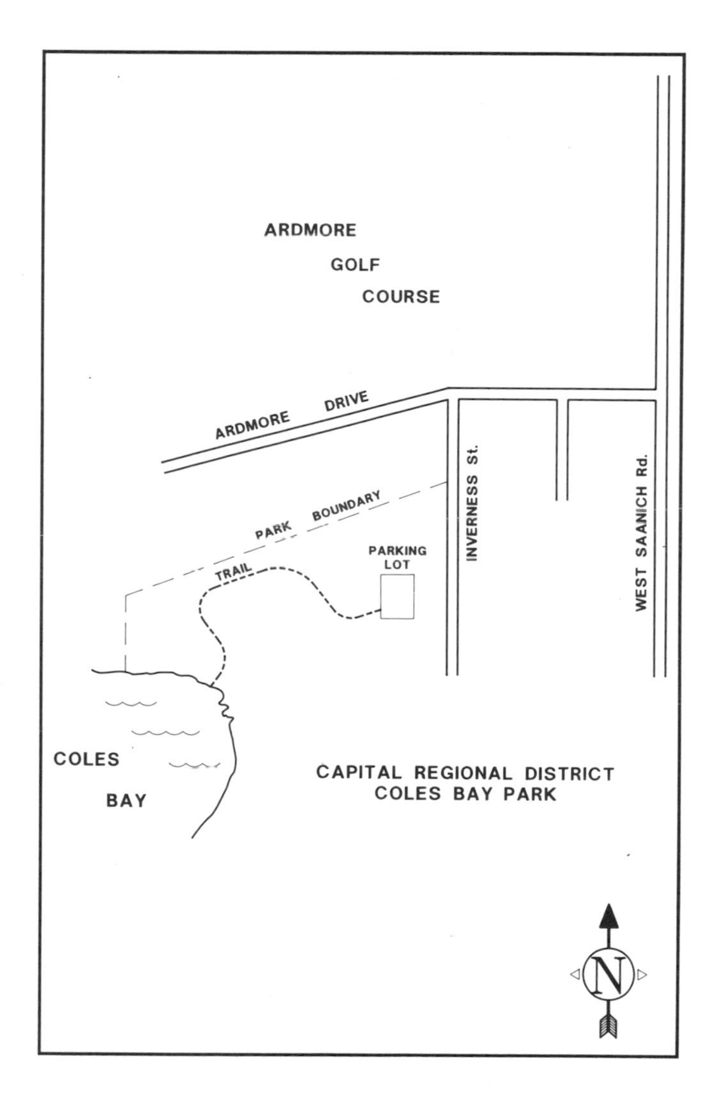

The Capital Regional District park at Coles Bay is reached from West Saanich Road. Turn west onto Ardmore Drive, then drive south on Inverness Street to the parking lot. Walk down one of the trails to the shore of Coles Bay.

The "salt and pepper" appearance of the Early Jurassic Saanich Granodiorite is well displayed by the rocks at the Capital Regional District park at Coles Bay. The dark minerals are hornblende and mica, the light minerals feldspar and quartz. The large, rounded dark areas are possibly pieces of Upper Triassic Karmutsen Formation lava that became incorporated into the Jurassic magma when it was intruded into the crust about 169 million years ago.

Outcrops along the shore are of igneous rock called granodiorite, meaning that the mineral composition consists of specific proportions of sodium and potash feldspar, quartz, hornblende and biotite mica. This rock forms part of a large igneous intrusive mass called the *Saanich Granodiorite*, which underlies the central part of the Saanich Peninsula and belongs to a large suite of plutons occurring throughout Vancouver Island called the *Island Intrusions* (see page 24). Many of the erratic boulders of granodiorite commonly found throughout the Victoria area were probably picked up by glaciers from outcrops in this area (Locality 8).

Clean, unweathered outcrops are light-gray- and black-speckled, medium-grained rocks in which the minerals feldspar (white with flat, shiny crystal faces), quartz (watery gray), hornblende (black needles) and mica (black platy grains) can be seen readily with the unaided eye, but more easily with a magnifying lens. Some areas weather a pinkish to orange colour, indicating the presence of the potassium variety of feldspar called **orthoclase**. Rounded, dark-gray, fine-grained inclusions several centimetres in diameter, which are richer in dark minerals, are a common and characteristic feature of these rocks. They represent fragments of older

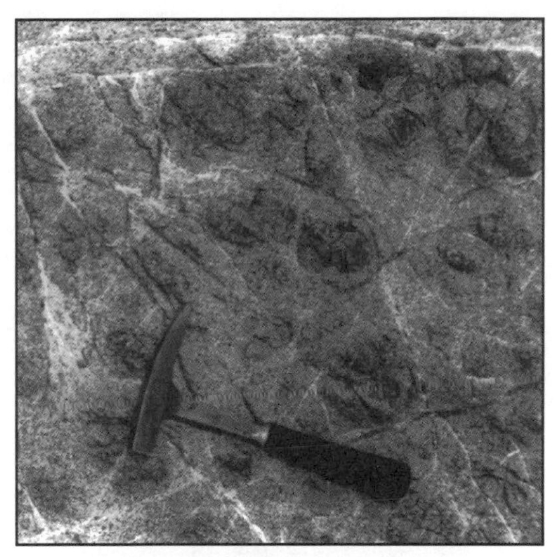

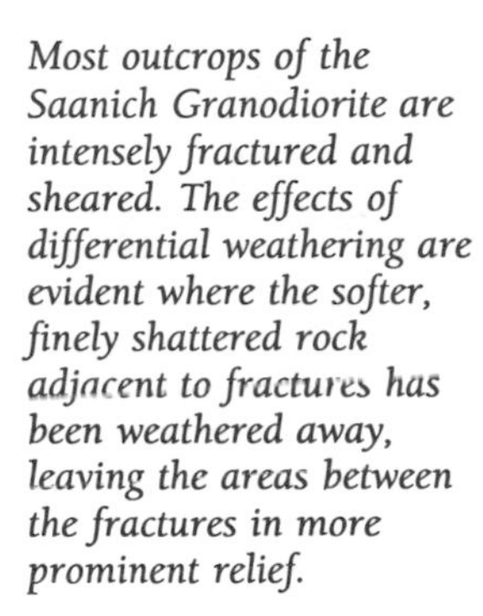

Most outcrops of the Saanich Granodiorite are intensely fractured and sheared. The effects of differential weathering are evident where the softer, finely shattered rock adjacent to fractures has been weathered away, leaving the areas between the fractures in more prominent relief.

Many of the homes on the slopes of Mt. Newton use the Saanich Granodiorite as the focus of attractive rock gardens.

rocks, probably pieces of the Upper Triassic *Karmutsen Formation*, which were entrained in the intruding magma when it was injected into the crust of the island from the hot mantle below. Narrow, irregular, grayish-white, very fine-grained dykes several centimetres wide cut the granodiorite at many places. Prominent at this locality are the many fractures, faults and crushed zones

South of Coles Bay, the beach below the bottom of Senanus Drive displays the Saanich Granodiorite where it has been fractured, or jointed, into vertical layers and cut by a small fault.

cutting through the granodiorite. Most trend, or **strike**, between 010° and 035° and incline, or **dip**, steeply to the east. Other outcrops of the *Saanich Granodiorite* are found east of Roberts Point, near the dock at the foot of James Island Road, near Brentwood Bay at the foot of Senanus Drive, at Ten Mile Point (Locality 7) and in many impressive rock gardens on Mt. Newton.

The *Saanich Granodiorite* has been radiometrically dated as 169+17 million years old. The means by which this age is determined is through the use of radioactive isotopes of the element potassium, abundant in the mineral hornblende, which is a common constituent of these plutons. The element potassium occurs as two isotopes: potassium 39 and potassium 40. Potassium 40 is radioactive and decays to the element argon 40 at a known rate, such that a gram of potassium 40 will decay to approximately half a gram of argon 40 in about 1.25 billion years. This means that measurable quantities of argon 40 occur in rocks ranging in age from a few thousand years to more than four billion years. The physics of the decay process are well known; thus by measuring the proportions of potassium 40 and argon 40 in the crystal structure of hornblende from the *Saanich Granodiorite*, it has been determined that the magma forming the pluton cooled about 169 million years ago, during the middle part of the Jurassic Period.

7.
TEN MILE POINT

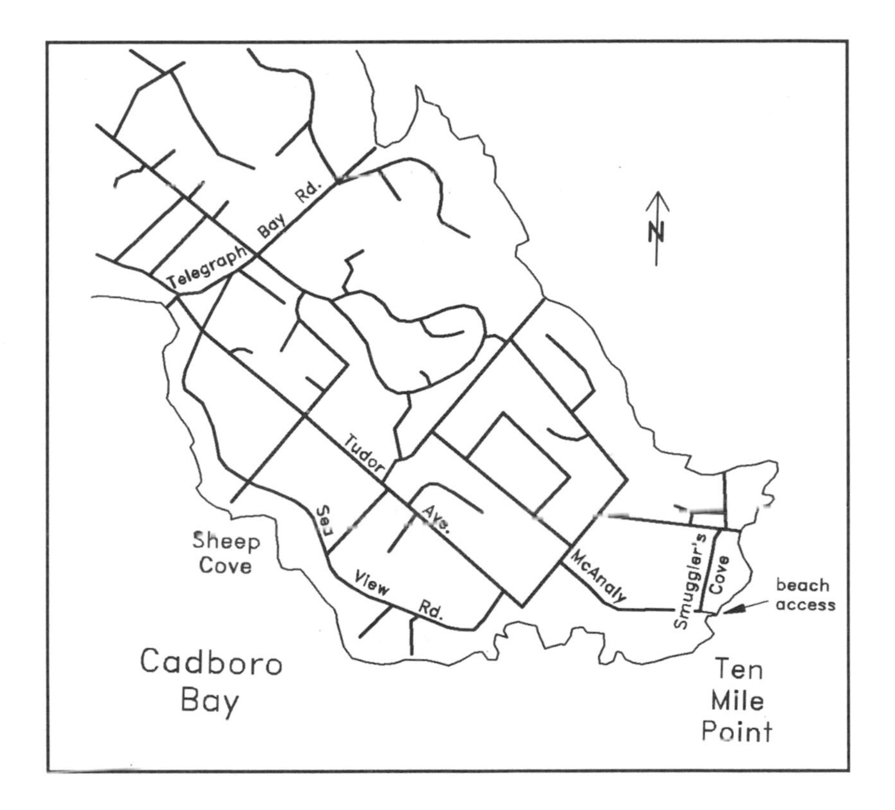

The beach access to Ten Mile Point is at the end of McAnally Road at its intersection with Smugglers Cove Road. It is best to visit this locality at times of low tide.

Like Coles Bay (Locality 6), Ten Mile Point and the entire peninsula is composed of igneous rocks assigned to the early Jurassic *Island Intrusions*. The rocks are mainly medium-gray- and black-speckled granodiorite cut by white quartz/feldspar dykes and veins. Most of the outcrops are covered by algae and lichens, giving them a dark-gray appearance. In several places the granodiorite is a rusty-orange colour, reflecting the presence of the feldspar called orthoclase. Other minerals include horn-blende, appearing as prismatic crystals, and the less common

The rocky coast of Ten Mile Point is formed from the Saanich Granodiorite.

Thin, light-gray dykes of quartz and feldspar cut the darker gray-weathering granodiorite at Ten Mile Point.

flakes of biotite mica. At a few places the rocks are a darker gray colour and may have the mineral composition of gabbro. Unlike outcrops of the *Saanich Granodiorite* at Coles Bay, Mt. Newton and Bear Hill, these rocks contain virtually no inclusions of the rocks into which these magmas were intruded.

The igneous rocks of the *Island Intrusions* form much of the rugged peaks of the Vancouver Island Ranges (see page 10). They occur as a series of large plutons, which form such mountains as The Golden Hinde, Elkhorn Mountain and Mt. Victoria. These rocks, formed from molten magma injected into the crust, are the coarse-grained equivalents of fine-grained rocks of identical chemical and mineralogical composition, which cooled from the same magma, but reached the earth's surface as volcanic lavas. These lavas occur throughout much of Vancouver Island and are called the *Bonanza Group*. In the Greater Victoria area, rocks uncertainly assigned to the *Bonanza Group* occur on the east shore of Squally Reach, south of Brentwood Bay in Saanich Inlet. The difference in crystal size between volcanic and plutonic intrusive rocks is related to the rates at which the magmas cooled. For magmas intruded into the crust where they cool deep beneath the surface, the cooling time is many millions of years, during which the crystallizing minerals grow to significant sizes. Lavas erupted at the surface, on the other hand, cool very rapidly, some so fast that no crystals form, the result being obsidian or volcanic glass. Thus it was that the magmas of the *Island Intrusions* were intruded into the crust some two hundred million years ago, where they cooled and from where they were later uplifted and exposed through erosion at the surface.

8.
McNEILL BAY, GONZALES BAY AND THE GONZALES OBSERVATORY

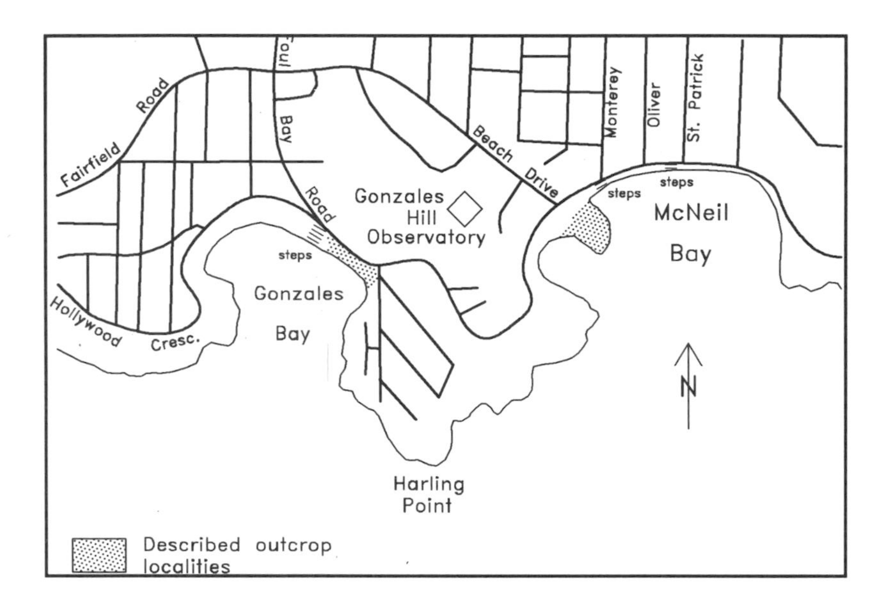

McNeill and Gonzales bays are separated from one another by Harling Point, from where there is an excellent view of Trial Island and the connecting waters of Haro Strait and the Strait of Juan de Fuca. Across the latter to the south is Port Angeles, on the north shore of the Olympic Peninsula of Washington, and to the east the several islands of the American San Juan Islands.

The locality in McNeill Bay is situated along the west side of the bay near the intersection of Beach Drive and Monterey Avenue. Walk down the steps across from the foot of Oliver Street and along the gravel beach toward the rocky ridge shown in the photograph. In the low cliff at the back of the beach, brown sandy clay overlies a more sandy and gravelly material. The brown clay is part of the *Victoria Clay*, laid down at the end of *Fraser Glaciation* when relative sea level was higher than at present (see page 35). When first exposed, this clay would have been blue-gray in colour and very soft, but it has since dried and weathered to the present brown colour. The more sandy and

A large glacial erratic (four metres in diameter) of Saanich Granodiorite rests upon Colquitz Gneiss in Gonzales Bay.

gravelly material underlying the clay may have been formed by slumping of blocks of stagnant glacial ice. In places, each of these materials contains marine shells, indicating that deposition took place in salt or brackish water.

The rocky ridge is best exposed at low tide and consists of contorted, interbedded, dark-gray **argillite**, schist and lighter-coloured, weakly metamorphosed siltstone of the *Leech River Complex*. Nodular masses and beds of chert are common, the latter commonly deformed into complex small folds. The rocks show well-developed glacial grooves and other features of glaciation.

The beach access to Gonzales Bay at the foot of Foul Bay Road provides excellent outcrops of the *Leech River Complex* where it occurs beneath the *Trial Island Fault*. Immediately southeast of the stairs and across the concrete abutment, dark-gray to black argillite and schist are interbedded with **radiolarian chert**. Radiolaria are microscopic planktonic creatures whose skeletons are composed of silica. The skeletons of these creatures make up substantial portions of modern deep-sea cherts and are important fossils for determining the age of such deposits. If you have a magnifying lens, you might see these tiny fossils embedded in freshly broken pieces of the hard, light-gray cherts, which weather a rusty-orange colour. Look for pin-head-sized, spherical inclusions that are slightly darker gray than the surrounding chert.

The *Trial Island Fault* is a thrust fault and appears to be a continuation of the *Survey Mountain Fault* (see Locality 9 and page 40) on which the *Leech River Complex* of the *Pacific Rim Terrane* was thrust beneath *Wrangellia*, here represented by the *Wark and Colquitz Gneiss Complex*. The surface trace of the fault occurs just above these outcrops, in the grassy and lawn-covered slopes above, but is nowhere well exposed. Above the fault the *Wark Gneiss* forms Gonzales Hill, upon which the *Gonzales Ob-*

A narrow rocky point within McNeill Bay exposes rocks of the Leech River Complex (top). Light-coloured siltstone is interbedded with argillite (middle), and, nearby, the strata are complexly folded as shown by contorted layers of chert enclosed by argillite (bottom).

Along the shore of Gonzales Bay, close to the beach access steps, is massive, black argillite of the Leech River Complex (above). Nearby, these argillites are interbedded with rusty-weathering radiolarian chert (below). The skeletons of tiny radiolaria can be seen with a magnifying lens as pin-head-sized spheres within the chert.

servatory is located. Note the large glacial erratic of *Saanich Granodiorite* resting upon *Colquitz Gneiss*.

The *Gonzales Heights Meterological Observatory* was designed by Francis Napier Denison, Victoria's first seismologist. From 1898 until the building's completion in 1914, earthquakes were recorded on a seismograph located in the Victoria Customs House on the corner of Wharf and Broughton streets. Following transfer of the seismograph to the new observatory, Denison continued to operate this and additional seismographs until his retirement in 1936. Shortly thereafter a shuffle in government departments transferred

The Gonzales Heights Meteorological Observatory atop Gonzales Hill was the site of Victoria's seismological observatory between 1914 and 1939.

responsibility for the seismographs from the Meteorological Service to the Dominion Observatory of the federal Department of Mines and Resources. In 1939 the seismograph station made a third move, this time to the Dominion Astrophysical Observatory on Little Saanich Mountain (see Locality 1). Currently there is one vertical, short-period seismograph operating in the *Gonzales Observatory*, from where data are telemetred to the Pacific Geoscience Centre near Sidney, which, since 1978, is western Canada's primary earthquake-recording and seismological institution.

9. GOLDSTREAM PARK, MT. FINLAYSON AND THE MALAHAT

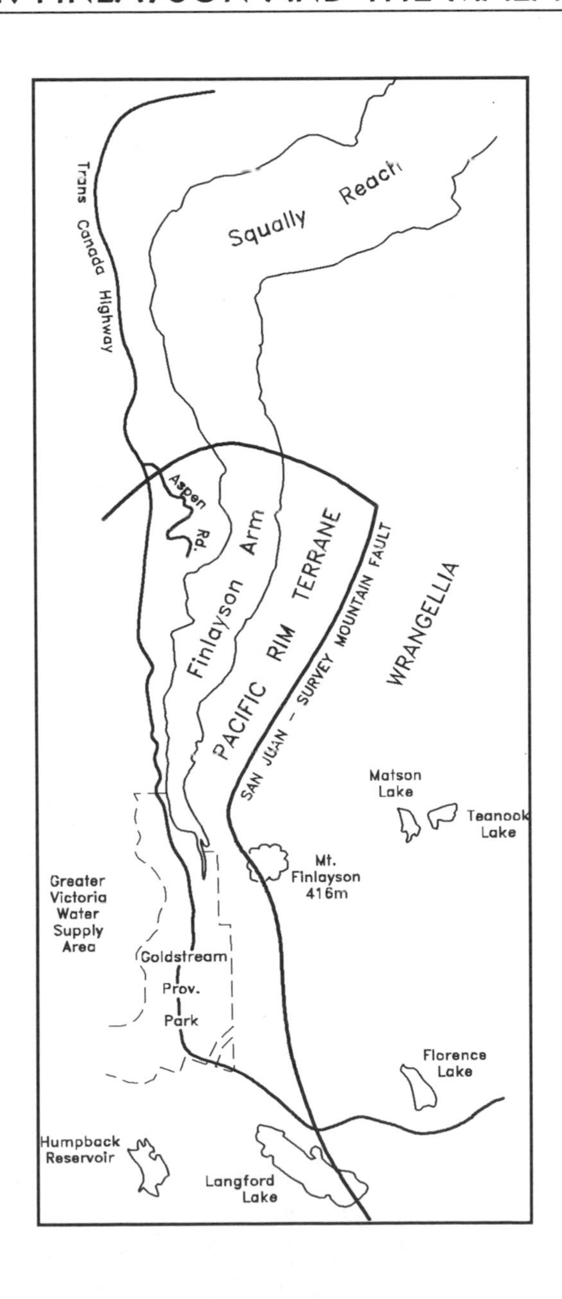

Geological relationships among the rocks of this area are complex, for it is in this region that you can observe the effects of collision between the *Pacific Rim Terrane* and *Wrangellia*. The *Pacific Rim Terrane* is here represented by metamorphic strata of the *Leech River Complex*, which have been emplaced beneath the *Wark Gneiss* at the southern edge of *Wrangellia* along the *San Juan – Survey Mountain Fault*, the surface trace of which crosses the middle slopes of Mt. Finlayson.

Roadside outcrops along the Trans-Canada Highway west of Helmcken Road consist of massive metamorphosed igneous rocks of the *Wark Gneiss*, one of the principal rock units of *Wrangellia* in the Victoria area. Shortly east of its intersection (Goldstream Avenue and Highway 1A), the Trans-Canada Highway crosses the surface trace of the *San Juan – Survey Mountain Fault* separating *Wrangellia* from the *Pacific Rim Terrane*. Beyond the intersection, roadside exposures of the *Pacific Rim Terrane* form part of the *Leech River Complex*, which, to near the turnoff on Aspen Road, consists of a nearly vertically inclined succession of metamorphosed and unmetamorphosed strata about 6 km thick. Near the intersection with Aspen Road the highway again crosses the *San Juan – Survey Mountain Fault*, bringing you back into the *Wark Gneiss* of *Wrangellia*.

The *Leech River Complex* is composed of sedimentary and igneous rocks which have sustained a low to moderate degree of metamorphism. Argillite, sandstone, chert and volcanic rocks of various kinds were deformed and metamorphosed to schists, gneisses and slates perhaps as much as eighty-five million years ago, following which, some thirty million years later, they were transported northward and thrust beneath the southern and western margins of *Wrangellia*. About forty-two million years ago, a second terrane, the *Crescent Terrane*, consisting of deep sea-floor volcanic rocks, was rammed beneath the *Pacific Rim Terrane* along the *Leech River Fault*, thus causing the metamorphic rocks of the latter to be compressed and tightly folded like an accordian against the backstop of *Wrangellia*.

A good place to see rocks of the *Leech River Complex* is along Goldstream River near the entrance to the Goldstream Park Campground, just off Sooke Lake Road. Rock exposures occur near the bridge over Goldstream River just past the gate into the campsite. Looking down from the walkway on the north side of the bridge, the strong uniform foliation in the streambed outcrops is parallel to the bridge and almost seems to be part of the bridge structure. Under the footings of the bridge the rocks are light grayish-green volcanic strata, forming a layer 3 m thick within a mass of dark gray to black slate. These rocks are readily accessible

Along Malahat Drive the Leech River Complex is exposed as a north-dipping succession of folded strata more than 5.8 km thick. The rocks consist of variably metamorphosed argilite, sandstone, chert and volcanic rocks that were folded into tightly compressed anticlines and synclines when, as part of the Pacific Rim Terrane, they were shoved beneath Wrangellia along the San Juan – Survey Mountain Fault. The dark-gray strata (right) are intruded by pale greenish-gray volcanic rocks (left).

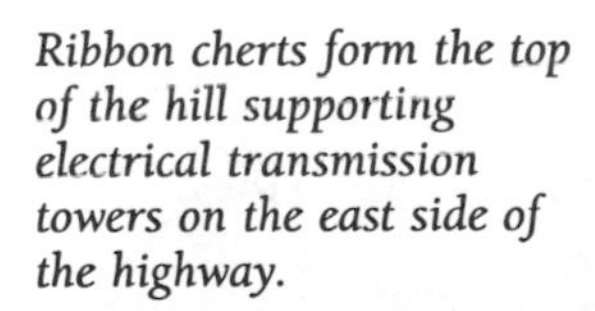

Ribbon cherts form the top of the hill supporting electrical transmission towers on the east side of the highway.

From the top of the hill supporting electrical transmission towers, the view to the south shows the estuary of Goldstream River where it enters Finlayson Arm. The surface trace of San Juan – Survey Mountain Fault traverses the middle slopes and the base of the prominent cliff on Mt. Finlayson on the east side of the estuary.

upstream (south) from the bridge by following one of the paths shown on the map. The foliation trends, or strikes, 125 to 135° and dips 80° to the northeast. The foliation, or **cleavage**, is a product of metamorphism and is formed by the parallel orientation of very fine flakes of white mica (**muscovite**), which causes the rocks to split into thin sheets and which gives the cleavage surfaces a lustrous sheen. Lenses of white quartz occur in clusters at several places within the dark-gray slate.

As you drive northward out of the park, through Goldstream Canyon and up the Malahat, you pass by steeply inclined, dark argillite, schist and minor thin **sandstone** beds and volcanic strata. Up the hill these rocks become progressively less metamorphosed, and thick layers of sandstone become more common than argillite. Near the transmission towers on the top of a hill on the east side of the highway, the rocks consist of approximately equal proportions of sandstone, argillite and chert, with lesser amounts of pillow basaltic lava. Farther north, fresh roadcuts expose a 150-metre-thick layer of pale grayish-green volcanic rocks which have been intruded into sandstones. Close to the Aspen Road turnoff, sandstones again become dominant. A short interval of grassy slopes covering the surface trace of the *San Juan – Survey Mountain Fault* separates the sandstones of the *Leech River Complex* from the *Wark Gneiss*.

From Aspen Road, the surface trace of the *San Juan – Survey Mountain Fault* crosses Finlayson Arm, then turns southward along the west side of the Gowlland Range to cross the west shoulder of Mt. Finlayson above Goldstream Park. The lower part of the mountain, as far as the base of the main cliffs at an elevation of 190 m, consists of *Leech River Complex* schist. At the base of the cliffs is the first of three strands of the *San Juan – Survey Mountain Fault*. The two other strands occur within the lower cliffs where, at an elevation of 270 m, the upper strand separates *Leech River Complex* schists below the fault from *Wark Gneiss* above the fault, the latter forming the main peak of the mountain.

From its intersection with Aspen Road, continue up the Malahat for 9.3 km (5.8 miles) to the second viewpoint overlooking Saanich Inlet and the Saanich Peninsula. The viewpoint is located directly above the abandoned limestone quarry at Bamberton.

Saanich Inlet and its southward continuation, Finlayson Arm, is a glacial fiord, probably carved during multiple glacial advances of the Wisconsinan stage of Pleistocene glaciation. Until its retreat some thirteen thousand years ago, ice occupied the fiord and supplied meltwater to the Goldstream drainage region via southeasterly flowing channels cut across the surface of the *Colwood Delta* (Locality 19). The inlet has average and maximum

From the viewpoint on the Malahat overlooking the abandoned cement plant at Bamberton, the northern part of the Saanich Peninsula is seen on the east side of the glacial fiord of Saanich Inlet. The prominent hill on the peninsula is Mt. Newton, formed of 170-million-year-old granodiorite. To the north (left) of Mt. Newton is Coles Bay (Locality 6).

Granitic rocks on the west side of the viewpoint are probably of Paleozoic age. The light-gray colour is lime dust from the cement plant below the viewpoint.

Mt. Finlayson rises 416 metres above Goldstream Park. At the base of the cliffs in the upper part of the photo, the San Juan – Survey Mountain Fault separates Jurassic and Cretaceous Leech River Complex metamorphic rocks of the Pacific Rim Terrane in the lower part of the mountain from Lower Jurassic metamorphic rocks of the Wark Gneiss forming the upper part.

depths of 120 m and 236 m respectively. A bedrock sill at the north end restricts water circulation, resulting in poorly oxygenated bottom waters in the deeper parts of the inlet. Recent studies of the bottom sediments in the inlet have identified massive layers of silt and clay which may have been transported into deep water by the action of submarine landslides, possibly induced by earthquakes.

Across the inlet is the pastoral Saanich Peninsula. The peninsula is home to important dairy farms, orchards, sheep farms, horse-riding stables and hobby farms of all kinds, developed mainly upon glacial sediments. The broad, prominent hill of Mt. Newton is formed from Early Jurassic *Saanich Granodiorite*, as is the smaller Bear Hill to the south. Beyond the peninsula the cliffs of James Island are visable, as are the hummocky uplands of San Juan Island, underlain by Triassic volcanic and sedimentary strata. In the far distance is the dormant Quaternary volcano, Mt. Baker.

At this viewpoint, the granitic rocks behind you are thought to be of Paleozoic age, perhaps of the same age as those at Finlayson Point (Locality 4). Their white to light-gray colour is due to a cover of lime dust left over from the abandoned cement plant below. The limestone here, at the Butchart Gardens across the inlet and at several other widely scattered localities throughout the region is of uncertain Paleozoic or possibly Triassic age.

10. ARMSTRONG POINT / ALLBAY ROAD

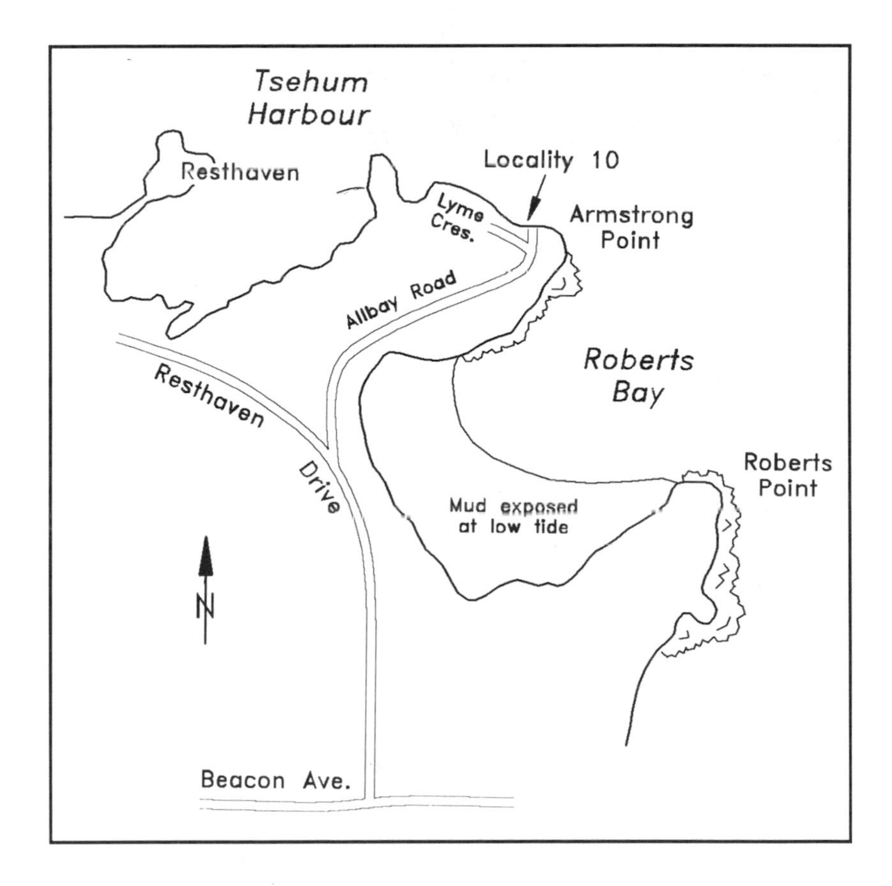

To reach this locality near Sidney, drive north on Resthaven Drive, then follow Allbay Road to its terminus close to the shore at Armstrong Point. This locality is not accessible at high tide.

This locality shows excellent examples of sedimentary rocks: sandstone, **siltstone**, shale and **conglomerate**, as well as many features associated with such rocks. The strata belong to the *Comox Formation*, the lowermost of the nine formations making up the *Nanaimo Group* of Cretaceous age. These strata rest upon volcanic rocks of uncertain, but possibly Paleozoic, age, the contact between the volcanics and the sediments marking what is

Cretaceous strata rest upon igneous rocks at many localities in the northernmost Saanich Peninsula.

At Armstrong Point at the end of Allbay Road, steeply inclined strata of sandstone, conglomerate and shale of the Comox Formation of the Nanaimo Group rest unconformably upon volcanic rocks, the latter possibly of Paleozoic age

The conglomerates consist of rounded pebbles and cobbles derived from Wrangellian Vancouver Island.

A bed of conglomerate at the left is overlain by steeply inclined siltstone and shale.

called an **unconformity**, or a gap, in the geological record .

Walk down the path and steps from the end of Allbay Road and, looking west (to your left), you will see excellent examples of steeply inclined strata consisting of sandstone, siltstone, shale and conglomerate. The bedding, or stratification, trends, or strikes, uniformly at 115° and dips 65 to 70° toward the north. The top, or youngest, part of the rock succession is toward the sea. Two sets of vertical fractures, or joints, cross the stratification trending 120° and 140°. On the outer, northernmost parts of the sea cliffs the rocks consist of gray siltstone with interbeds of gray sandstone 100 to 200 millimetres thick. Round **ironstone concretions** a few centimetres in diameter are common. Toward the south and lower in the stratigraphic succession, the proportion of sandstone to siltstone increases and beds of conglomerate are present. One prominent bed of conglomerate, less than a metre thick, contains round pebbles of chert, quartz, brick-red jasper, dark-green volcanic rocks and granitic rocks.

The unconformity is an irregular surface separating the sandstone and conglomerate of the *Comox Formation* from the dark-green, massive, structureless volcanic rocks forming the low cliffs near the heads of the bays and furthest from the shore. The volcanic rocks are also dislocated by many closely spaced, randomly oriented joints. The volcanic rocks are of uncertain Paleozoic age and, in late Cretaceous time, some eighty million years ago, formed the bedrock surface upon which the sedimentary strata of the *Comox Formation* accumulated in much the same way that sand and gravel of modern beaches are being deposited today. At the time they accumulated, the strata were approximately horizontal, although the surface upon which they were deposited was quite irregular. Since that time the rocks and the unconformity have been tilted to their present, moderately steep inclination by folding and faulting as a consequence of the collision of the *Pacific Rim* and *Crescent terranes* with *Wrangellia.*

11.
WITTY'S LAGOON PARK

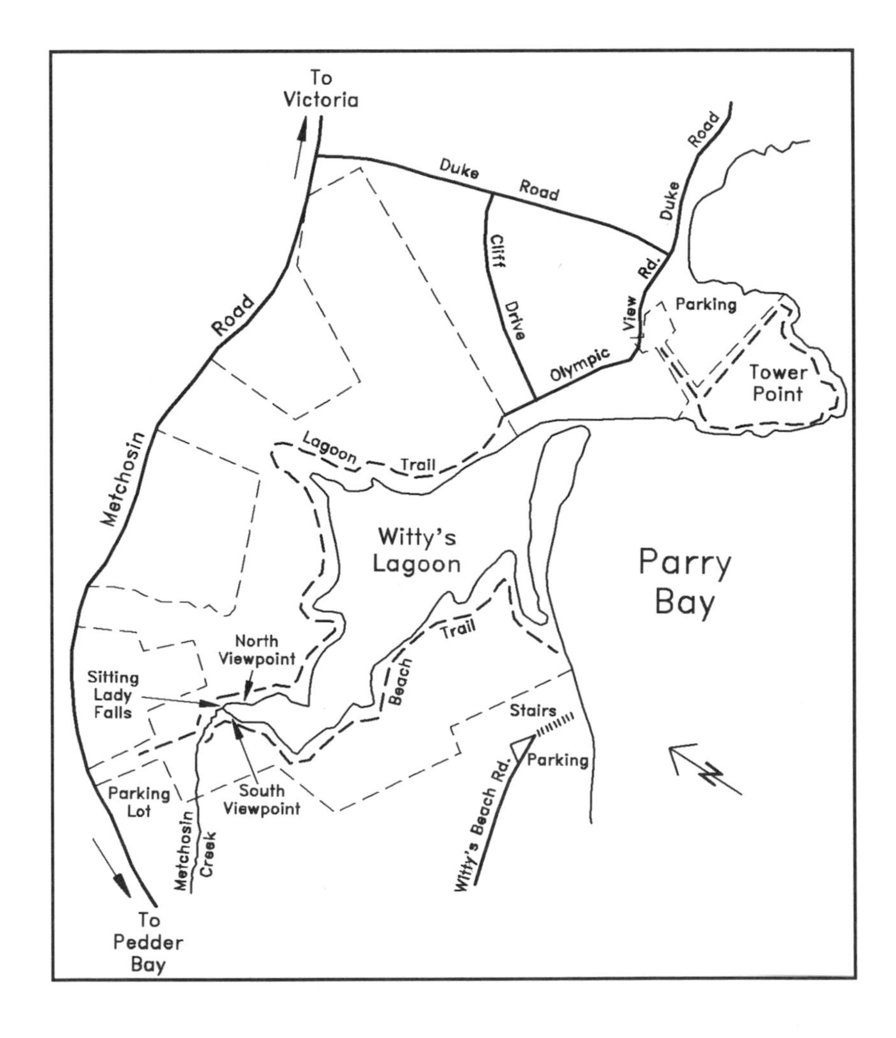

The Capital Regional District park at Witty's Lagoon is an excellent place to see features of the *Crescent Terrane*, represented by the *Metchosin Igneous Complex* which forms the south end of Vancouver Island south of the *Leech River Fault*. The main area of the park can be reached by trails from the parking lot off Metchosin Road. A separate area of the park at Tower Point is accessible from a parking area off Olympic View Road, west of its intersection with Duke Road.

Metchosin Creek cascades over pillow basalt of the Metchosin Igneous Complex at Sitting Lady Falls in Witty's Lagoon Park.

From the parking lot at the main park walk down the trail, keeping left, to the north viewpoint indicated on the map, where you will get a view of Sitting Lady Falls. Bedrock exposed at the falls is dark gray-green pillow basalt. From the viewpoint, an indistinct stratification of the lava flows, with a low inclination to the northwest, can be seen in the cliff face. The cliff forming the falls probably coincides with a northwest-trending, steep fault that is marked by prominent fractures and a narrow zone of sheared and crushed basalt close to the base of the cliff. The pillow structures give the outcrops a rounded, lumpy surface, reflecting the shapes of the pillows, which are as much as a metre in length. The pillows can be seen in the upper part of the rock face just below the south viewpoint.

From the trail to the south viewpoint, you can look down into the bed of Metchosin Creek, above Sitting Lady Falls, where you will see what appear to be boulders up to a metre in diameter. These, in fact, are lava pillows, still part of the intact bedrock. Weathering and stream erosion have removed the softer material between the pillows so that you get a good three-dimensional view of typical pillows.

From the south viewpoint, you have a good view of Witty's Lagoon, consisting of extensive mudflats through which Metchosin Creek meanders to the sea. The lagoon is formed within a channel eroded by Metchosin Creek between nine thousand and five thousand years ago, when the shoreline was as much as 10 m lower than at present (see page 35). Between the base of Sitting

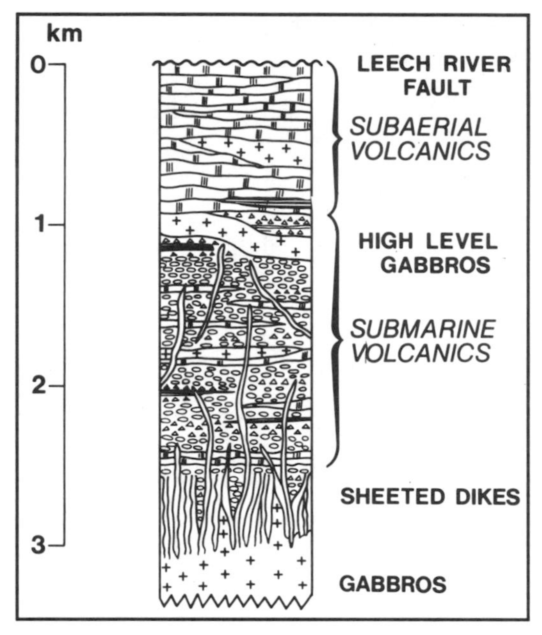

The stratigraphic succession of the kinds of rocks making up the Metchosin Igneous Complex (Crescent Terrane) is about three kilometres thick. At the base are layered gabbros, which pass upward into sheeted dykes that were injected into the spreading ridge where the complex formed. Above the dykes are submarine pillow basalts that built a pile upwards toward the sea surface. These are capped by lava flows that built a volcanic island not unlike modern Iceland. At the top of the succession is the Leech River Fault separating the Crescent Terrane from the Pacific Rim Terrane.

Lady Falls and the shoreline, the creek would have cut its channel down to meet the lowered sea level, but, as sea level rose, that portion of the channel was flooded to form the present shallow lagoon. Sand and gravel eroded from sea cliffs south of the park were transported northward by **longshore currents** to form the spit and sand bars exposed at low tide.

Additional excellent exposures of pillow lavas occur at Tower Point. On the point and sea cliffs, the characteristic features of the pillow basalts are well displayed in clean outcrops above the high-tide line. These dark-green, fine-grained rocks commonly contain amygdules filled with quartz and calcite, which appear as white spots up to a centimetre in diameter. Several vertical, green, **vesicular** dykes, up to a metre wide, trend across the point, and a minor east-dipping fault is exposed on the western side of the point. Several outcrops display piles of basalt pillows with flattened bases and shapes indicative of their having been squeezed together while the lava was still hot and plastic. Conspicuous, light-gray to almost white, obviously erratic boulders of granodiorite can be seen lying on the surface of the pillow basalts.

The *Metchosin Igneous Complex* is thought to have developed as an oceanic island, not unlike Iceland, about fifty-four million years ago. The pillow basalts which are exposed here, as well as at Sooke Potholes Provincial Park and at many other localities throughout the Sooke and Metchosin region, are only part of the

Pillow breccia forms outcrops along the shore of Tower Point.

Excellent examples of pristine pillow basalt can be seen along the shores of Tower Point and nearby islands.

complex, the remainder consisting of large intrusions and dykes of gabbro, the coarse-grained equivalent of basalt, and thick lava flows which erupted both beneath and above sea level. During the course of eruption, the lava flows commonly broke up into blocks, called **breccia**, or fine dust, called **tuff**. These broken pieces of lava, tuff and volcanic glass accumulated in layered deposits between the pillowed and massive flows and in the spaces between the pillows.

Many of the pillows seen in this area contain abundant round, white amygdules that are commonly arranged in layers close to the margins of the pillows. These amygdules were originally **vesicles** that have been filled by crystals of calcite or other minerals. Vesicles form when gas, dissolved in molten lava, separates from the melt, causing it to froth. If the pressure due to the weight of the overlying water is sufficiently great, the gas does not separate and no vesicles form. Thus there is a rough correlation between the depth below sea level at which the lava erupted and the vesicularity of the lava; with increasing depth the degree

From Tower Point the view of the estuary of Metchosin Creek shows broad tidal flats bordered by pillow basalt of the Metchosin Igneous Complex.

of vesicularity decreases. From this relationship we can conclude that the pillow basalts of the *Metchosin Igneous Complex* erupted in moderately deep to shallow water, but not as deep as the present Pacific Ocean spreading ridges.

The coarser-grained components of the *Metchosin Igneous Complex*, which formed the deeper layers of oceanic crust, consist of large intrusions of gabbro and complex arrays of gabbro dykes called **sheeted dykes**. These occur along the shores of Becher Bay and East Sooke Park (Localities 12 & 13).

The total thickness of the *Metchosin Igneous Complex* is about 3 km, and because these oceanic rocks occur on land, the total complex is called an **ophiolite**. Ophiolites are fragments of oceanic crust that have become incorporated into continents.

The *Metchosin Igneous Complex* forms only part of the *Crescent Terrane*. Other parts include the *Crescent Volcanics*, which form the magnificent Olympic Mountains across the Strait of Juan de Fuca, and several other volcanic formations of what is called the *Coast Range Basalt Province*, extending from the Leech River Fault to northern California.

12. EAST SOOKE PARK: AYLARD FARM

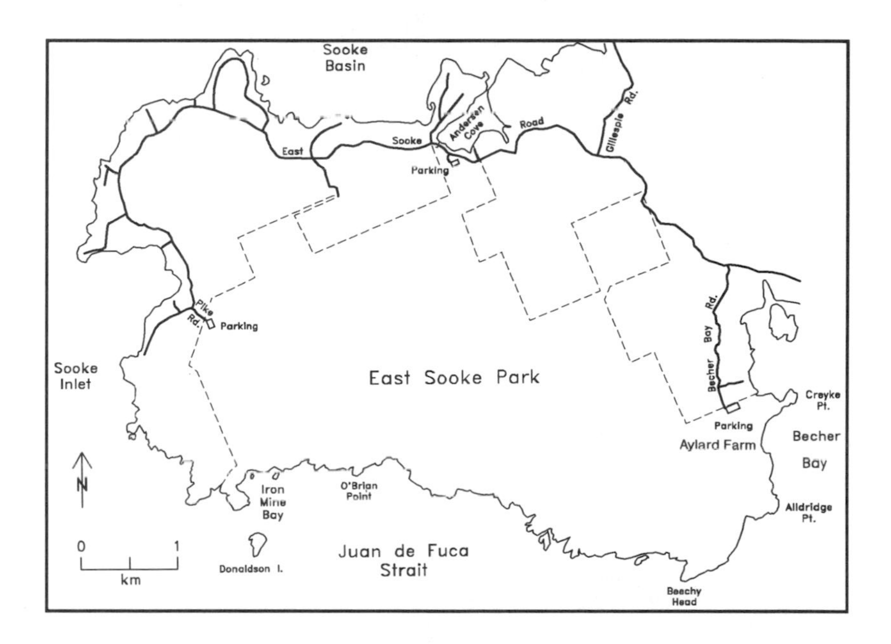

From Victoria, East Sooke Park is reached via the Trans-Canada Highway, the Island Highway, Metchosin Road, Happy Valley Road, Rocky Point Road and East Sooke Road. At the intersection of East Sooke and Becher Bay roads, turn south on the latter and proceed to the parking lot at Aylard Farm, East Sooke Park.

East Sooke Park is underlain by components of the *Metchosin Igneous Complex* of the *Crescent Terrane*, different from those seen at Tower Point of Witty's Lagoon Park (Locality 11). In addition, the park displays sandstone and conglomerate strata of the Sooke *Formation*, part of the early Tertiary *Carmanah Group* (see page 31).

From the parking lot, walk across the lovely meadows of the park to Creyke Point, on the south shore of Campbell Cove in Becher Bay. The prominent dark, rocky bluffs are composed of a geometrically complex array of gabbro dykes forming a sheeted dyke complex. To most people, including most geologists, these outcrops look like massive, structureless lava; however, the dykes

From Creyke Point, the view southward along the west shore of Becher Bay toward Alldridge Point displays rock of the Metchosin Igneous Complex and the overlaying Sooke Formation, both of Early Tertiary age.

In the middle distance, the rocky beach consists of boulder conglomerate of the Sooke Formation, which formed in an environment similar to that of the modern beach.

Boulders of gabbro (hammer head) are tightly cemented by sandy pebble-conglomerate (hammer handle).

can be discerned on close inspection by their **chilled margins**. When magma is injected into fissures in rock, it quickly begins to cool and crystallize. The molten lava will cool most quickly, and therefore form the finest crystals, at the margins where it is in contact with cold, previously solidified rock. This fine-grained skin is called a chilled margin. Here at Creyke Point, the dykes have been deformed such that their chilled margins are bent and twisted and difficult to see with an untrained eye.

The origin of sheeted dykes is linked to oceanic spreading ridges, where oceanic crust is formed through the process of sea-floor spreading. The dykes are injected along a narrow zone at the axis of

The Sooke Formation here consists of a lower, cross-bedded sandstone layer and overlaying boulder conglomerate, the latter forming the bouldery beach shown here.

The sandstone is finely cross-bedded, as shown by the inclined strata of this photograph.

the ridge as a consequence of separation of the two adjacent plates. The pillow basalts overlying the dykes form when lava is quenched as it extrudes onto the ocean floor. With each succeeding injection the previously accreted dykes move apart, forming a thick layer of dykes underlying pillow lavas. The lengths of the dykes of the *Metchosin Igneous Complex* are estimated to range between 400 and 700 m, and their widths between 3 and 10 m.

The walk southward along the shore to Alldridge Point takes you across a stretch of twenty-five-million-year-old beach. The rocks along the shore consist of sedimentary strata of the *Sooke Formation*, which accumulated on the eroded surface of the *Metchosin Igneous Complex*. The upper part of the *Sooke Formation* exposed on the beach consists of large boulders of gabbro, some as much as 4.5 m in diameter, enclosed within a matrix of granule and pebble sandstone and conglomerate. These large blocks look much like glacial erratics, but careful examination

Veins of quartz and feldspar cut through the gabbro (top), which, at Alldridge Point, displays a seal petroglyph (left).

shows them to be embedded in a matrix of conglomerate and sandstone. Much like the present coast, the upper part of the *Sooke Formation* appears to have developed in a shallow marine coastal environment, dominated by rocky headlands, sandy coves and sea stacks, where frequent storms generated large waves that crashed upon the coast. That these blocks have not travelled far is shown by their angular outlines. Perhaps they fell from steep sea cliffs onto a sand and gravel beach that was pounded by storm waves. In other words, the coast and beach of twenty-five million years ago was probably much the same as it is today.

The lower part of the *Sooke Formation* is exposed at several places in the low cliffs backing the beach. It consists of orange to rusty-weathering, massive and locally **cross-bedded** sandstone, interbedded with and overlain by boulder conglomerate. No fossils have been recovered from the formation at this locality, but elsewhere it has yielded marine fossils, indicating that the sediments accumulated during the Oligocene stage of the Tertiary Period, some twenty-five million years ago.

At Alldridge Point, the *Metchosin Igneous Complex* is represented by coarsely crystalline gabbro, the deepest layer of Tertiary oceanic

Gabbro of the Metchosin Igneous Complex consists of dark plagioclase-feldspar and large crystals (up to five centimetres in length) of pyroxene, the latter seen brightly reflecting sunlight beside the penny.

crust exposed in the complex. With increased depth in the ophiolite, the finer-grained sheeted dykes pass downwards into gabbroic rocks of composition similar to the dykes and lava flows, but with much coarser crystals, which can be seen easily without a magnifying lens. These gabbros form much of the peninsula and occur at several other localities throughout the region underlain by the complex. In general, the gabbros are fairly homogeneous in appearance, but in some places a vague layering can be seen, resulting from differences in crystal sizes or mineral composition. The gabbros consist mainly of gray, calcium-plagioclase feldspar, dark greenish pyroxene crystals up to five centimetres in length and small reddish crystals of **olivine**. The gabbro is cut by many white veins and dyklets composed of fine- to medium-grained quartz and feldspar. The gabbros here display Native petroglyphs, representing a salmon and a sea lion.

13. EAST SOOKE PARK: IRON MINE BAY

Iron Mine Bay is located at the western end of East Sooke Park (see map page 119). Follow East Sooke Road for approximately 10 km past its intersection with Becher Bay Road to its junction with Pike Road. The parking lot is located at the end of Pike Road, from where an easy twenty-minute walk through magnificent forest brings you to the beach.

Rock exposures along the shore of the bay and Pike Point consist of fine-grained basalt and sheeted gabbro dykes, which have been intruded by a mass of coarse-grained gabbro called the *East Sooke Gabbro*. During the last stages of crystallization of the gabbro intrusion, the remaining molten fluid, depleted of most of its magnesium, iron and calcium minerals, was squeezed out of the gabbro body and injected into the surrounding fine-grained rocks. The high pressures of this process caused the surrounding rocks to fracture, allowing the molten fluid to penetrate along the cracks and crystallize as light-coloured quartz diorite, a rock consisting of fine-grained feldspar and quartz. The resulting mix of broken surrounding rock and quartz diorite is called an **agmatite**.

Copper mineralization in this area was first discovered by Jeremiah Nagle in 1863 in sea cliffs near O'Brien Point, about 1.5 km east of Iron Mine Bay, where Iron Mine Road meets the Coast Trail. This discovery led to the sinking of one of the earliest shafts in bedrock in British Columbia. Several other deposits have since been found and developed during the early 1900s, including the Willow Grouse on the northeast slope of Mt. Maguire and the Copper King, marked by a plaque on the Iron Mine Trail. These two deposits contain the copper mineral **chalcopyrite**, which occurs with the mineral hornblende along fractures in the rock. These mines shipped small quantities of ore between 1915 and 1918.

Fog enshrouds the rocky coast of Iron Mine Bay.

Coarse-grained gabbro of the East Sooke Gabbro (hammer head) has intruded fine-grained basalt and sheeted dykes (lower hammer handle).

Light-coloured quartz diorite has been intruded along fractures in dark-coloured gabbro, resulting in a fragmented mixture called an agmatite.

14. FRENCH BEACH PROVINCIAL PARK

French Beach Provincial Park is located off Highway 14, 21 km west of the centre of the town of Sooke (intersection of Highway 14 and Otter Point Road). The parking lot is close to the beach, which is most accessible at low tide. From the entrance onto the beach, outcrops of the Tertiary *Sooke Formation* occur approximately a half-kilometre to the northwest (right).

Before reaching the sandstone outcrops, you must cross a patch of large boulders of pillow basalt of the *Metchosin Igneous Complex*. These are **lag** boulders and gravels, left behind by longshore currents that have removed the finer sandy particles and distributed them along the shoreface of the beach. The presence of these large basalt boulders indicates that bedrock of the *Metchosin Igneous Complex* (*Crescent Terrane*) is close by.

The outcrops of the *Sooke Formation* consist of horizontally stratified, pale- to rusty-brown-weathering, thin-bedded sandstone, pebbly sandstone and conglomerate. The sandstones consist mostly of quartz, basalt fragments and lesser amounts of feldspar and other minerals such as **ilmenite** and **magnetite**, all of which are cemented together by **calcite**. The sandstones also contain numerous whole and broken fossil shells of clams and snails. In some places the fossil debris is so concentrated that the deposit is called a **coquina**. The coquinas of the *Sooke Formation* suggest that the formation accumulated in shallow marine waters not unlike the environment that forms the modern beaches along the Strait of Juan de Fuca.

The broad, low, flat surface formed by the bedrock and extending seaward from the low sea cliffs is an excellent example of a **wave-cut terrace**. These features result from erosion due to the spray and splash of storm waves, which cut the sea cliffs landward, forming the terrace, the surface of which is just below high-tide level.

Southeast of the main trail entrance onto the beach and just north of where a bridge crosses a creek, you will observe a ridge formed by a relict sand dune composed of sand eroded from cliffs of *Sooke Formation* sandstone. The ridge probably developed close to the end of *Fraser Glaciation* during the period of lowering of relative sea level.

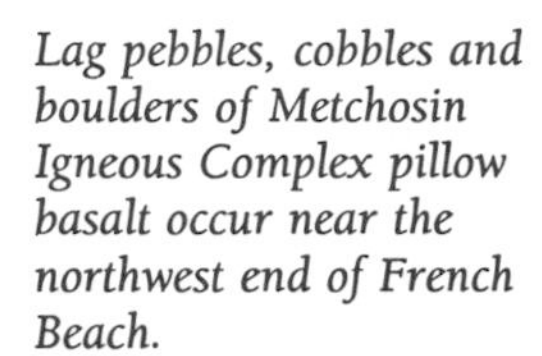

Lag pebbles, cobbles and boulders of Metchosin Igneous Complex pillow basalt occur near the northwest end of French Beach.

Fossil clams are common in the sandstones of the Tertiary Sooke Formation. The left side of the scale is in inches and the right side in centimetres.

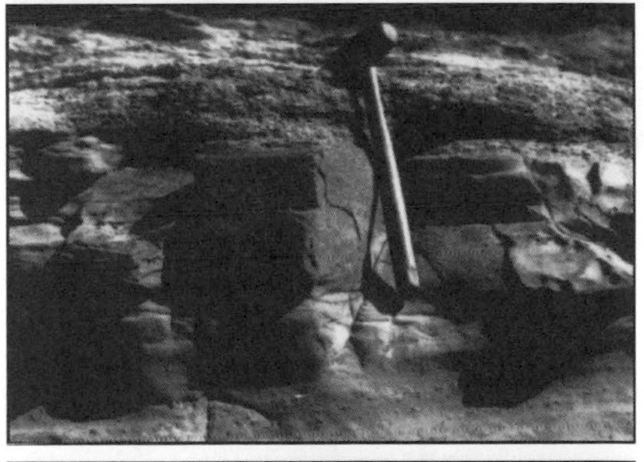

Sandstone beds of the Sooke Formation are capped by a layer of shell coquina (hammer head).

The wave-cut terrace extending seaward from the low sea cliffs at the northwest end of French Beach has been formed by the erosive action of storm-wave spray and splash on sandstone strata of the Sooke Formation.

15. CHINA BEACH PROVINCIAL PARK

The entrance to China Beach Provincial Park is located off Highway 14 about 36 km west of the centre of the town of Sooke (intersection of Highway 14 and Otter Point Road). From the parking lot, the trail to the beach takes you on a fifteen-minute walk through a magnificent forest of Douglas fir, hemlock, alder and arbutus. Outcrops of the Tertiary *Sooke Formation*, similar to those at French Beach Provincial Park (Locality 14), occur along the beach to the northwest (right).

The imposing cliffs consist of gently inclined, pebbly sandstone strata of the *Sooke Formation*, which, in the far distance to the northwest, overlie dark-coloured volcanic rocks of the *Metchosin Igneous Complex* (*Crescent Terrane*). In addition to a storm-wave-cut cave at the base of the cliffs, the sandstones display horizontal potholes, carved by cobbles and pebbles that originally were constituents of the formation, but which were eroded out by storm waves, thus serving to carve the holes. To the east (left) of the cliffs, the base of the formation consists of boulder conglomerate in which large, angular black boulders embedded in the pebbly sandstones consist of volcanic rocks of the Metchosin Igneous Complex. This conglomerate is similar to the beach conglomerate at Aylard Farm at East Sooke Park (Locality 12).

At the northwest end of the beach, coarse pebble and boulder conglomerate forms the lower part of the Sooke Formation.

The imposing cliffs at the northwest end of China Beach are composed of fossil-bearing sandstone of the upper part of the Tertiary Sooke Formation.

In the photo at right, horizontal potholes have been eroded into the vertical face of the sandstones by the combined action of storm waves and pebbles; the latter, loosened from the sandstone, act as abrasives to further excavate the holes.

16. LOSS CREEK PROVINCIAL PARK

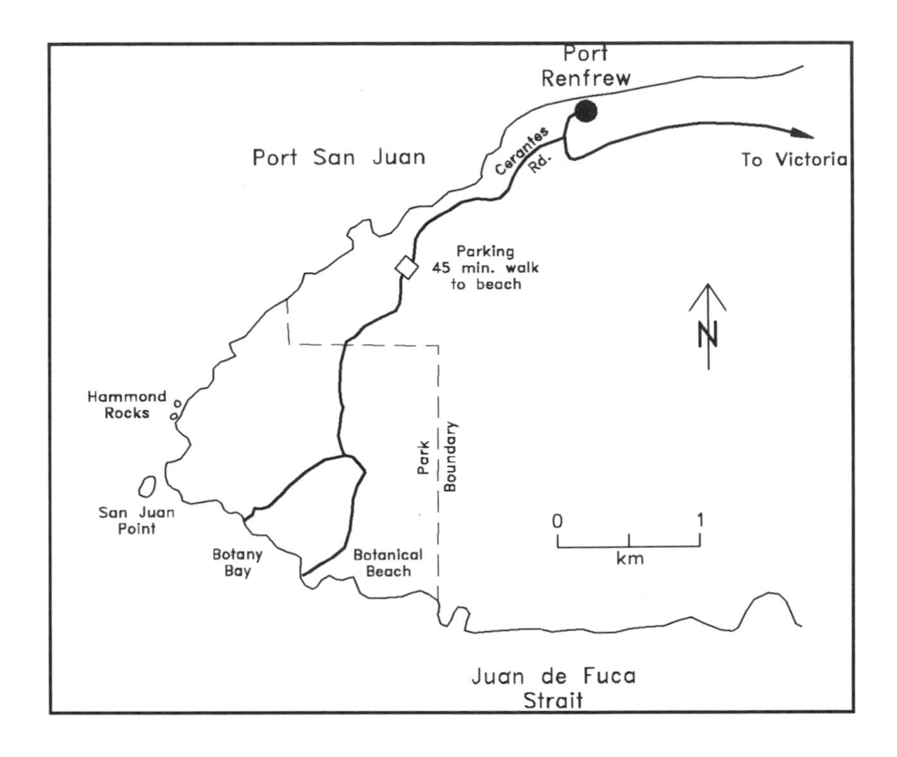

Loss Creek Provincial Park is situated at the base of a steep hill on Highway 14, some 52 km west of the centre of the town of Sooke (intersection of Highway 14 and Otter Point Road).

On your drive westward from Sooke to Loss Creek Provincial Park, you have been passing by outcrops of the *Metchosin Igneous Complex* of the *Crescent Terrane*, as well as a few exposures of the *Sooke Formation*. A short distance past the entrance to China Beach Provincial Park (Locality 15), a prominent roadside outcrop exposes coarse fragments of volcanic rocks, originally part of the *Metchosin Igneous Complex* but which have been eroded from their original location to form a basalt- and gabbro-pebble conglomerate belonging to the *Sooke Formation*. As you drive down the hill towards Loss Creek, you pass intensely shattered and sheared volcanic rocks of the *Metchosin Igneous Complex*. Upon crossing the bridge over Loss Creek, the next

Roadside outcrops on the south side of Loss Creek consist of intensely fractured and sheared massive volcanic rocks of the Metchosin Igneous Complex of the Crescent Terrane.

These rocks were deformed when they were shoved against and beneath the Pacific Rim Terrane (Leech River Complex) along the Leech River Fault, which is colinear with the valley of Loss Creek. Leech River Complex rocks occur as greenish-gray **chlorite** *schist across the bridge over Loss Creek.*

At Sombrio Point on the Strait of Juan de Fuca, the Leech River Fault passes seaward between the low, rocky point of Metchosin Igneous Complex volcanic rocks (Crescent Terrane) in the far distance and the narrow ridge of Leech River Complex metamorphic rocks (Pacific Rim Terrane) in the middle distance.

outcrop you see consists of greenish-gray schist of the *Leech River Complex* of the *Pacific Rim Terrane*. Thus from one side of the creek valley to the other, you cross the *Leech River Fault* separating the *Crescent Terrane* on the south from the *Pacific Rim Terrane* to the north. Westward from the park, the *Leech River Fault* continues to Sombrio Point, from where it extends offshore beneath the Strait of Juan de Fuca.

The oblique westerly focused aerial photograph and associated drawing show the gun-barrel-straight valley containing Bear and Diversion reservoirs and Loss Creek, all of which occur along the Leech River Fault. To the south (left) of the fault, rocks of the Crescent Terrane (Metchosin Igneous Complex) have been shoved beneath those to the north (right), which consist of Leech River Complex strata. In the distance are the Strait of Juan de Fuca and the Olympic Mountains of northern Washington, with Cape Flattery in the far distance to the right.

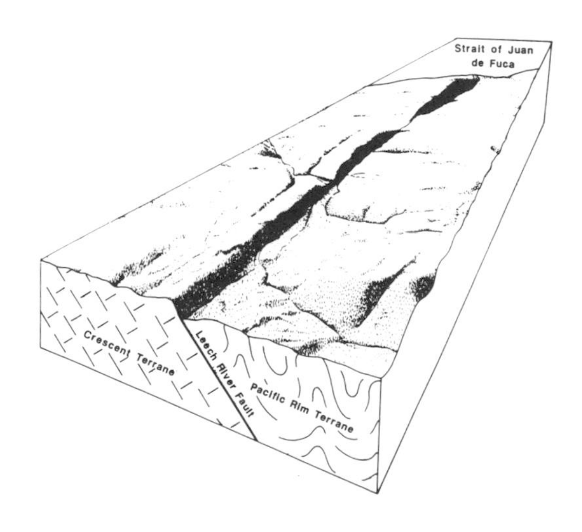

17. BOTANICAL BEACH PROVINCIAL PARK

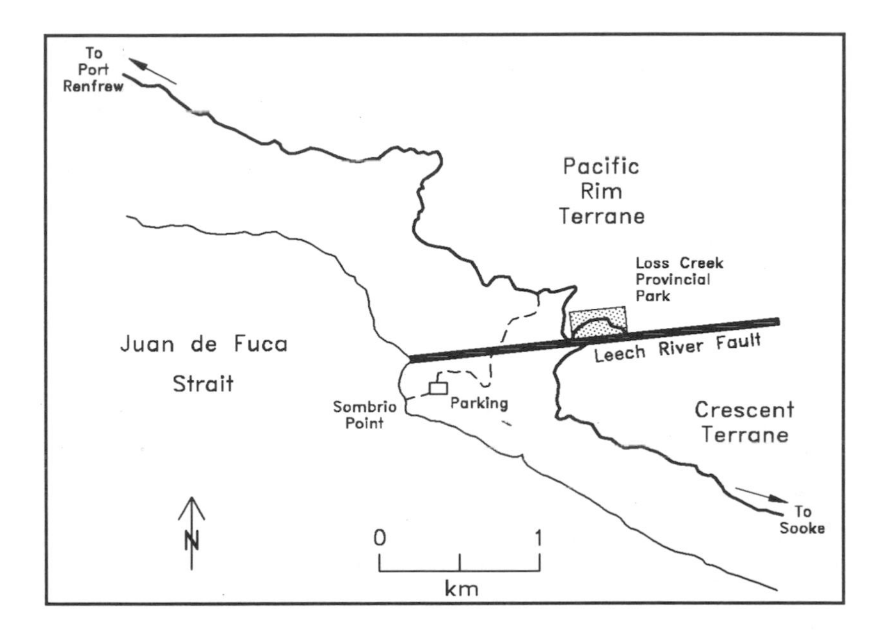

From Loss Creek, the highway to Port Renfrew lies entirely within the Pacific Rim Ter*rane*, roadside outcrops consisting of dark, platy schist of the *Leech River Complex.* As you approach Port Renfrew, the degree of metamorphism of these rocks progressively lessens until near the town, where the rocks consist of black argillite and, less commonly, sandstone.

To get to Botanical Beach and adjoining Botany Bay, both in Botanical Beach Provincial Park, follow Highway 14 through Port Renfrew, then turn left on Cerantes Road and proceed for 3 km along a rough gravel road to the main trailhead, some forty-five minutes' walk from the beach. Those with high-clearance, four-wheel-drive vehicles can proceed further to the last parking lot, from where the walk to the beach takes about ten minutes. At the time of writing, plans were being developed to improve this road.

Botanical Beach Provincial Park is spectacular, both biologically and geologically. In addition to the extraordinary abundance and

Horizontal strata of the Tertiary Sooke Formation rest with angular unconformity upon tilted slabs of argillite and dense, fine-grained sandstone of the Jurassic and Cretaceous Leech River Formation (upper). In the lower photograph, coarse-grained Sooke Formation sandstone is preserved in fractures separating blocks of dense Leech River sandstone.

diversity of intertidal marine life, the area displays a wide variety of geological features between the two trail ends at Botanical Beach and Botany Bay. Close to where the trail exits onto Botanical Beach, you can see an excellent example of an **angular unconformity** separating the folded and tilted argillites and dense, fine-grained sandstones of the Jurassic and Cretaceous *Leech River Complex* from the relatively horizontal, thin, coarse-grained to pebbly sandstone of the Tertiary *Sooke Formation*. The story told by the rocks here begins some 135 million years or more ago, when the muds and sands that ultimately became argillite and sandstone of the *Leech River Complex* accumulated at or near the toe of a continental slope, perhaps somewhere in the vicinity of the San Juan Islands. Some eighty million years later, these sediments were transported northwestward as the *Pacific Rim Terrane* and rammed beneath Vancouver Island (*Wrangellia*) along the *San Juan – Survey Mountain Fault* (see page 40). During this process the argillites and sandstones of the *Pacific Rim Terrane* (*Leech River Complex*) were intensely folded as they were

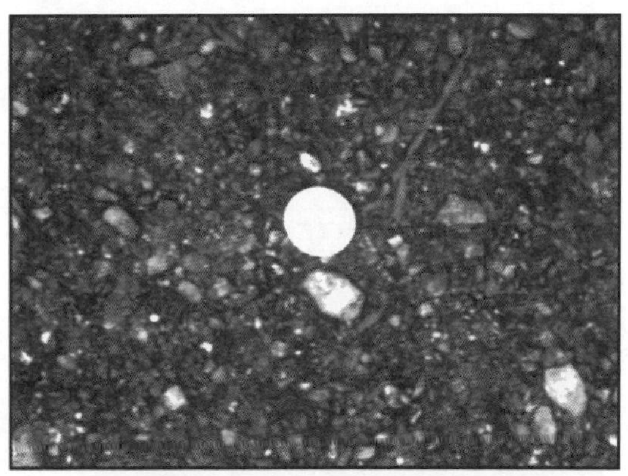

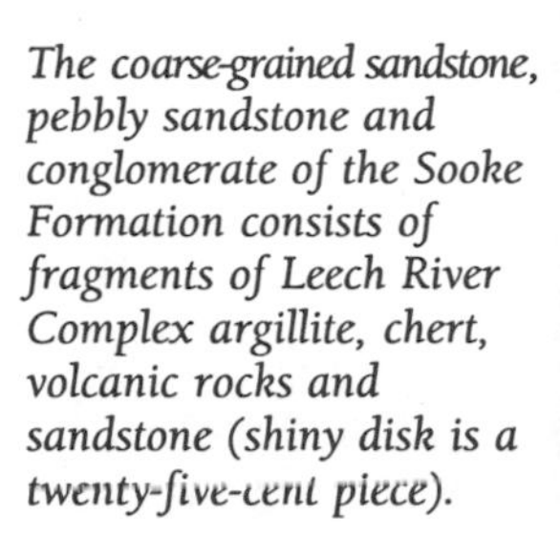

The coarse-grained sandstone, pebbly sandstone and conglomerate of the Sooke Formation consists of fragments of Leech River Complex argillite, chert, volcanic rocks and sandstone (shiny disk is a twenty-five-cent piece).

Prominent sea cliffs of Sooke Formation sandstone form the point separating Botanical Beach from Botany Bay.

compressed against the backstop of *Wrangellia*. Following accretion, the folded argillites and sandstones were eroded such that the products of erosion, namely fragments of argillite, sandstone and a variety of other rock types from other sources, were deposited as the *Sooke Formation* upon the bevelled, upturned edges of *Leech River Complex* strata. In some places, the coarse sandstones of the *Sooke Formation* are preserved in open fractures and crevices in Leech River sandstones. The *Sooke Formation* is about twenty-five million years old; thus, in terms of the missing rock record between the two formations, the angular unconformity here represents an interval of some 110 million years.

Moving northwestward around the point between Botanical Beach and Botany Bay, you see spectacular sea cliffs and a broad wave-cut terrace formed from the *Sooke Formation*. The effects of **differential erosion** are clearly evident in the sea cliffs, where dark-gray, irregularly shaped masses of more wave-resistant

The effects of differential erosion are here displayed by slight differences in the amounts of intergranular cement holding the grains of sandstone together. The prominent, irregularly shaped masses of dark-gray sandstone are well cemented by calcite, whereas the surrounding pale, smoothly eroded sandstone is softer due to less intergranular cement.

The unique, uniformly sized, hemispherical depressions in the soft Sooke Formation sandstones were eroded by the hard spines of purple sea urchins. By these and other processes, the deep tidepools of the wave-cut platform are enlarged and deepened.

sandstone stand out in rugged relief as compared to surrounding smooth-looking, pale-brown and rusty sandstone. The difference is due to variations in the amount of calcite cementing the grains together, the pale-coloured strata having somewhat less cement than the dark-gray, irregularly shaped sandstone masses. Other features of note in the cliffs are large sea caves, as well as clusters of baseball- to football-sized sandstone **concretions**. The latter were formed shortly after the sandstone was deposited, through precipitation of calcium carbonate and iron oxide around a nucleus such as a fossil fragment embedded in the sandstone.

The ribbed, step-like appearance of the rocks of Botany Bay is due to northerly dipping argillite and sandstone strata of the Leech River Complex (above), which, locally, is cut by veins of white quartz (below).

A walk across the wave-cut terrace takes you beside many deep pools eroded into the horizontal and comparatively soft sandstone of the *Sooke Formation*. At several places you can see that purple sea urchins occupy circular depressions in the walls of the pools, such that a honey-comb effect is imparted to the sandstone. Sea urchins produce these depressions by eroding the soft sandstone with their hard spines, in this way enlarging the pools over time. Many marine animals, such as snails, are known to carve holes in rock, particularly in shoreline exposures of sandstone of the Cretaceous *Nanaimo Group* of the Gulf Islands; however, the extraordinary erosive power of sea urchins, which have populated the oceans for more than 450 million years, is yet to be appreciated by geologists.

Continuing northwestward around the point into Botany Bay, you walk across northerly inclined argillite and sandstone strata of the *Leech River Compex*. Locally, these strata enclose veins of white quartz. Similar veins in the vicinity of Valentine Mountain, west of Victoria, contain gold.

18. ISLAND VIEW BEACH AND COWICHAN HEAD

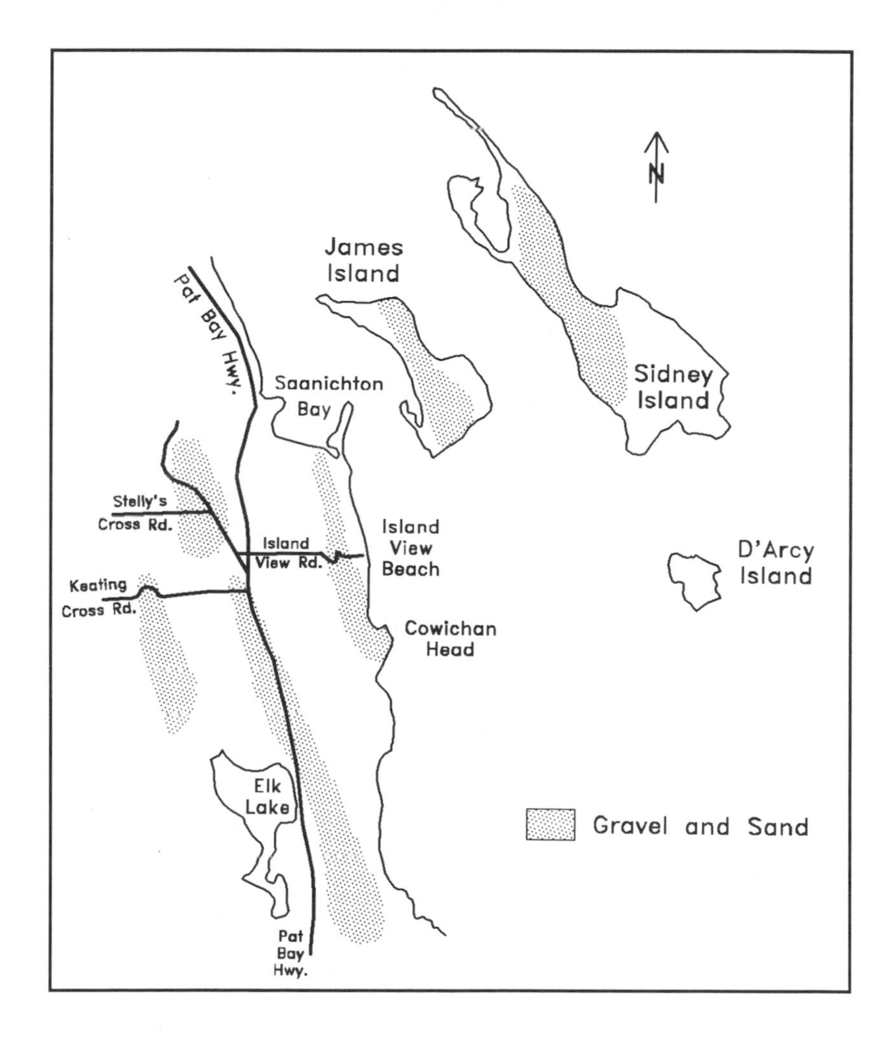

This locality is reached via the Pat Bay Highway and Island View Road. Eastward from the highway, Island View Road crosses a flat valley bottom underlain by black organic soil, then climbs over a low ridge trending parallel to the coast and extending from Cowichan Head to Saanichton Bay. This ridge is one of several nearly

Gravel is mined from the Quadra Formation at the Butler Brothers pit on Saanich Peninsula.

parallel ridges composed of unconsolidated glacial deposits. James Island and the northern two-thirds of Sidney Island form parallel ridges to the east. To the west is a ridge extending from Cordova Bay, past Elk and Beaver lakes to just north of Keating Cross Road. Other ridges to the west are crossed by Keating Cross Road and Stelly's Cross Road. All of these ridges consist principally of unconsolidated materials, much of which is advance outwash of the *Fraser Glaciation*, known as the *Quadra Formation*. In this area the *Quadra Formation* includes extensive gravel deposits, which accumulated in front of a glacier that advanced down the Cowichan Valley and overrode Saanich Peninsula before joining with the main body of ice advancing south along the Strait of Georgia. These gravels have been quarried from several pits on Keating Cross Road (Butler pit and Central Saanich pits), Cordova Bay Road (Trio Pit and Saanich Municipality pits), and at the south end of Central Saanich Road.

As the glaciers advanced, the *Quadra Formation* accumulated as a continuous outwash fan extending across Saanich Peninsula and James Island to at least as far as Sidney Island. The fan was eroded by the advancing glaciers and moulded to form the present ridged topography. *Vashon Till* (see page 36) overlies the older glacial and interglacial sediments, forming the drumlinoid ridges of which the natural cross-section at the south end of James Island is typical.

The exposures in the cliffs south of Island View Beach show something of the internal stratigraphy and structure of the *Quadra Formation*. The walk south along the beach to these exposures at Cowichan Head takes fifteen to twenty minutes. The beach at the parking lot is part of a gravel and sand spit, known

The broad drumlin form of James Island is evident in this aerial view. The sea cliffs at the south end of the island expose sands and gravels of the Quadra Formation. Southward from the cliffs are extensive shallows of sand and gravel, indicative of the former extent of the island, which has sustained considerable wave erosion and northward retreat of the south coast.

as Cordova Spit, which has been built northward by deposition of sand and gravel moved by longshore currents from the cliffs at Cowichan Head, a process similar to that by which Coburg Peninsula has been formed at Esquimalt Lagoon (Locality 19). The flat area between the beach and the hillslope is underlain by mud, which grades upward and shoreward into the modern grassy marsh. Underlying the mud is a fifty-centimetre-thick layer of peat, which accumulated in a freshwater marsh when the sea level was lower than at present. The contact between the peat and the overlying mud has been carbon-14-dated as 2,450 years old, indicating a rise of sea level since that time.

Further evidence of the recent rise of sea level can be seen on the beach at the north end of the cliffs at Cowichan Head, where stumps and roots of trees protrude from the beach about 1.5 m below high tide. Wood from these stumps has been carbon-14-dated as 2,040 years old.

Erosion of the cliffs occurs principally by landsliding. At their north end, where the sediments include substantial amounts of silt and clay, the material collapses in large, irregular masses, whereas at the south end, where the cliffs are composed principally of sands and gravels, the unconsolidated material comes down as individual particles due to a lack of cohesive clay, a process known as **ravelling**. The large boulders and cobbles on the beach, also fallen from the cliffs above, are **lag deposits**, left behind by the processes of longshore drift, which, during winter storms at high tide, have removed the finer material to construct Cordova Spit to the north. The boulders have a wide variety of

Viewed from the Beachcomber RV Park at the end of Campion Road, the sea cliffs of Cowichan Head extend northward along the shore of Haro Strait. The cliffs expose strata formed as the result of Fraser and earlier glaciations and include, in upward order, the Muir Point, Cowichan Head and Quadra Formations. The top and steepest part of the cliff is formed of the Vashon Till and the Victoria Clay, the latter a marine clay that is now forty metres above sea level as a result of the net effects of glacial rebound of the land and the postglacial rise in sea level.

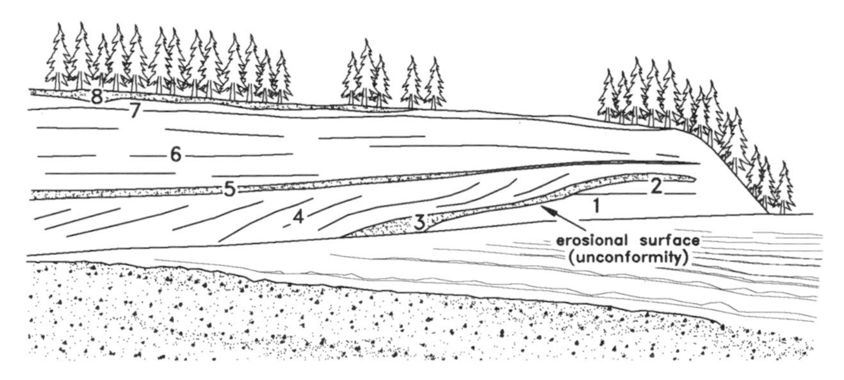

This drawing represents the stratigraphy of Cowichan Head as seen in the photo. A description of the numbered stratigraphic units is given in the text.

mineral compositions, indicative of their varied sources both on Vancouver Island and as far away as the Coast Mountains north of Vancouver. Recently, in an effort to retard the process of cliff erosion, a wall composed of very large blocks of quarried granodiorite has been placed at several places at the base of the cliffs.

From the beach facing Beachcomber RV Park south of Cowichan Head, the view to the north displays the cliffs as shown in the photo and accompanying diagram above. About halfway along the lower part of the cliff, you can see a distinctive layer of dark, silty clay, about 4 m thick, beginning just south of the erosion-retarding wall of quarried granodiorite, and labelled unit 3 on the diagram. This material is a layer of marine muds containing abundant organic material as well as shells and twigs and a layer of stones along its base. If you dig into it you will notice the unmistakable smell of estuarine mud. You can see how this layer rises northward and is draped over a ridge of older sandy and gravelly sediments of units 1 and 2. The material of unit 3, called the *Dashwood Drift*, reflects a rise in sea level immediately prior to the beginning of the *Olympia Interglacial Period*, which followed the so-called *Semiahmoo Glacial Period*, during which unit 3 was deposited. Unit 1 consists of about 10 m of light-gray sand, unit 2 of 12 m of brownish gravel and sand; together these are parts of the *Muir Point Formation*, which possibly accumulated during a pre-Wisconsinan interval of glaciation, perhaps more than eighty thousand years ago (see diagram on page 142).

The marine shells of unit 3 belong to species that presently occur off the Alaska coast, indicating a cool environment during the *Olympia Interglacial Period*. The remains of barnacles attached to stones collected from the unit have been identified as being those of the modern species *Balanus glandula*, which occurs in brackish estuarine environments between southern California and the Aleutian Islands.

The age of pieces of wood recovered from unit 3 are beyond the resolving capability of carbon 14 (practical limit = forty thousand years). Carbon from shelly material has yielded an age of about thirty-five thousand years, but this is not thought to be reliable. In any case, unit 3 is considered to represent either the last stage of *Semiahmoo Glaciation*, or the first phase of the *Olympia Interglacial Period*, the date being around sixty-two thousand years.

Units 4 and 5 have been designated respectively as the lower and upper members of the *Cowichan Head Formation*, a succession of silt, sand and gravel that accumulated during the *Olympia Interglacial Period*, a lengthy interval of warm climate preceding the *Fraser Glaciation*. Unit 4 consists of about 10 m of rusty-brown sand and gravel beds, inclined to the south and containing casts of marine worm tubes. Evidently these sands accumulated in shallow salt or brackish water following the rise in sea level evidenced by unit 3. Above unit 4, unit 5 consists of a prominent bed of silt and sand, some 7 m thick, containing fine particles of plant material and occasional pebbles. These sediments probably were deposited as flood plain silts from streams flowing across an emerging sea floor, during lowering sea levels heralding the onset of *Fraser Glaciation*. In places, at the contact between units 4 and 5, there is a reddish brown soil layer that formed shortly before the land was covered by the

Fossil shells of clams and snails are clearly seen in the marine clay of unit 3 near the south end of the quarried-granite retaining wall.

flood plain sediments. Plant fragments from the soil give an age of about 35,600 years, indicating that the soil formed during the *Olympia Interglacial Period.*

The flood plain silts of unit 5 form a prominent marker horizon which can be traced almost the full length of the cliffs. The way unit 5 dips to the south is related to the underlying ridge of sediments (units 1, 2 & 3) on which the sediments of the *Cowichan Head Formation* (units 4 & 5) were deposited. This dip continues southward so that these sediments disappear below the base of the cliff in that direction.

Above the *Cowichan Head Formation* is a thirty-metre-thick succession of stratified sand and gravel assigned to the *Quadra Sand* (unit 6), the advance outwash sand of the *Fraser Glaciation* discussed above. Measurements of the direction of inclination of cross-bedding within unit 6 indicate that the outwash streams from which these sediments were deposited flowed from the west, northwest, north and northeast. Radiocarbon ages ranging from about 24,400 to 28,800 years have been obtained from organic materials collected from the *Quadra Sand* at Comox on Vancouver Island and at Point Gray near Vancouver.

Looking at the south end of the cliff nearest Cowichan Head you will see, at the top, a nearly vertical cliff about 10 m thick. This is represented by units 7 and 8 in the diagram. Unit 7, ranging from 5 to 25 m thick, is the brown *Vashon Till* of the *Fraser*

Sand and gravel of the Quadra Formation forming the lower and middle slopes at Cowichan Head are overlain by cliff-forming Vashon Till and Victoria Clay.

Glaciation, dated at less than seventeen thousand years. Unit 8, about 2 m thick, is a brown- weathering glaciomarine stony clay, known locally as the *Victoria Clay*, which is probably no younger than about 12,750 years. From this distance the two units look very similar. The *Victoria Clay* and the *Vashon Till* support a nearly vertical face because they contain a significant amount of clay, which provides cohesive strength. The gravels, on the other hand, have little clay, thus ravel down to slopes of between 35 and 40°. At the south end of the cliffs the ravelled sand and gravel largely obscures units 5 and 4.

You will recall reading in Part One (see page 34) how the position of sea level during the Pleistocene Epoch was the result of a combination of lowering, due to withdrawal of water from the oceans to form the large ice sheets, and depression of the land surface due to the weight of the ice. These effects explain why the glaciomarine *Victoria Clay* occurs some forty metres above current sea level, at the top of the cliff. When the glaciers melted, vast amounts of water were returned to the sea, causing sea levels to rise by as much as 150 m. At the same time, but much more slowly, the land surface, which had been depressed by the weight of the ice by as much as 200 m or more, began to rise as the crust rebounded to its pre-glaciation level.

19. COLWOOD DELTA

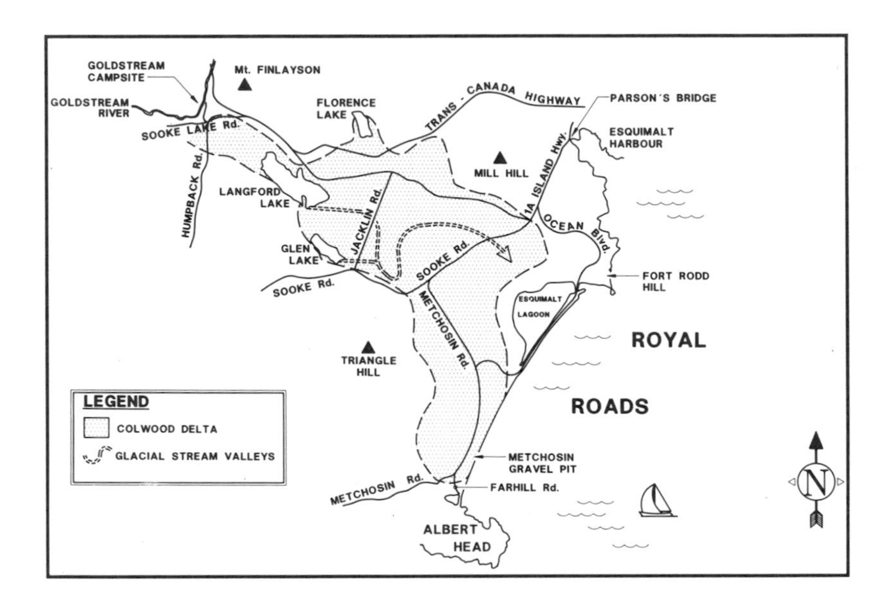

The *Colwood Delta* is a large gravel delta built by streams flowing from melting glacial ice west of Langford Lake and discharging into the sea north of Albert Head. The surface of the delta is 70 to 80 m above present sea level, approximately at the elevation of the shoreline when ice of the *Fraser Glaciation* was receding from the Victoria area about thirteen thousand years ago.

The following is a description of a route to be followed, with intermediate stops, illustrating the important features of the delta and enabling you to recognize and understand its land forms. From Parson's Bridge over Mill Stream, where it flows into the upper end of Esquimalt Harbour (see map on page 63), proceed southwest along the Island Highway (1A). Along this section of the route you can see numerous outcrops of *Wark Gneiss* mantled by a thin veneer of glacial sediments. The road climbs from near sea level at Parson's Bridge to about 60 m at Colwood Corners on the northern edge of the delta.

The map shows the delta to be irregular in shape, with arms extending to the south and northwest from near the junction of

The walls of the Metchosin Gravel Pit on Metchosin Road expose inclined foreset beds, overlain by topset outwash gravels, each deposited as part of an ice-retreat delta of the Fraser Glaciation, about 13,000 years ago (above and below).

Sooke and Metchosin roads. When the *Colwood Delta* was being built, about thirteen thousand years ago, the area between Esquimalt Harbour and Colwood Corners was covered with stagnant glacial ice; Saanich Inlet and Finlayson Arm were ice-filled, and meltwater from stagnant ice in the Goldstream watershed flowed eastward to the sea south of Esquimalt Lagoon.

Before reaching Colwood Corners, turn off to the southeast (left) on Ocean Boulevard toward Fort Rodd Hill Park. Instead of proceeding to the park, continue on down the hill and cross the narrow bridge onto Coburg Peninsula. About 100 m past the bridge, stop at the Royal Roads monument. Looking south, you can see the gentle curve of Coburg Peninsula, which has been built by wave-induced longshore currents carrying sand and gravel eroded from the front of the delta, northward to build the spit that now almost encloses Esquimalt Lagoon. The tides result in strong currents flowing in and out of the lagoon, which scour the channel beneath the bridge and keep the channel open.

Extending eastward and southeastward from Sombrio Point on the Strait of Juan de Fuca, beneath Glen Lake and the northern part of the delta, the *Leech River Fault* continues offshore at Esquimalt Lagoon and the Coburg Peninsula. Across the delta the fault has no surface expression, indicating a lack of activity since deposition of these sediments. Along its length the fault separates rocks of the *Metchosin Igneous Complex* of the *Crescent Terrane* on the south from *Leech River Complex* metamorphic rocks of the *Pacific Rim Terrane* to the north.

Looking across Esquimalt Lagoon, you will see that there are no bedrock outcrops along the southern two thirds of the shoreline, in contrast to the rocky shore along the northernmost part. From the lagoon the ground surface gently rises westward and southward to merge with the surface of the delta. Stagnant glacier ice, about 100 m thick, probably occupied the site of the lagoon while the delta was being built. This mass of stagnant ice, together with that between Parson's Bridge and Colwood Corners, diverted meltwater streams southward to deposit the sediments of the main delta lobe about 2 km farther south.

Proceed southward along Coburg Peninsula to the southern end of Esquimalt Lagoon where Lagoon Road comes in from the west. From this point you can see the large gravel pit that has been excavated in the delta front, as well as the upper surface of the delta, which, at this location, is about 75 m above sea level. As you proceed up Lagoon Road towards its junction with Metchosin Road, the soil exposed in the roadside ditches consists of brown and gray silty marine clay, called the *Victoria Clay*, which accumulated at the end of the *Fraser Glaciation*. As the road continues up through a gully eroded in the delta front, you will pass small exposures of medium to coarse gravels, with stratification dipping toward the sea. These are **foreset beds** of the delta and are inclined downwards in the direction that the river system flowed at the time they were laid down.

Turn left on Metchosin Road and proceed through the large gravel pit, which has been worked here since the early 1900s. In some exposures you may see inclined foreset beds of the delta overlain by horizontal **topset beds** deposited by glacial meltwater streams flowing to the sea, which stood 75 to 80 m above its present level.

Continue through the pit to Farhill Road, near the southern

Stagnant glacial ice, possibly one hundred metres thick, probably occupied Esquimalt Lagoon during construction of the Colwood Delta. Coburg Peninsula, enclosing the lagoon, has been built by northward-moving longshore currents carrying sand and gravel eroded from the front of the Colwood Delta.

limit of the delta. At the upper end of Farhill Road are several abandoned gravel pits. Farther down the road toward Albert Head, brown-weathered *Victoria Clay* is exposed in the roadside ditches. Farhill Road ends at a short gravel spit, which has been built across Albert Head Lagoon in much the same way as Coburg Peninsula was constructed across Esquimalt Lagoon. The south end of the spit abuts against the rocky shore of Albert Head, which is composed of pillow basalt of the *Metchosin Igneous Complex* (see page 31). The flow of tidal currents in and out of the lagoon is not strong enough to maintain an open channel. North of the spit, the steep, wooded bluffs mark the delta front, which has been eroded by storm waves since the time when sea level reached its present position.

Retrace your route along Farhill Road and Metchosin Road and stop at the north edge of the Metchosin Gravel Pit property. From there you can look southward over the pit operation and, to the west side of the private road, see a good example of a **kettle**, a closed depression formed when an isolated block of glacier ice becomes buried in the outwash gravels deposited by meltwater streams. As the ice melted, the gravels over and around the ice slumped inwards to form the depression. Numerous other kettles existed on the delta surface, but were removed in the course of gravel mining. The presence of these kettles confirms that there were numerous masses of stagnant ice in the area during the time when the delta was being constructed.

Continue along Metchosin Road, then turn westward (left) onto Sooke Road. This portion of the route is located on the gravel surface of the delta; small ridges and depressions mark channels eroded by streams flowing across this surface during delta construction. In this area, the route is along the contact between the *Colwood Delta* on the northeast and volcanic rocks of the *Metchosin Igneous Complex, small outcrops of which occur to the southwest.*

From its intersection with Sooke Road, the route northward (right) along Jacklin Road dips down into a low area, then rises a few metres to Belmont Secondary School. The low area is a meltwater channel crossing the road at right angles. About thirteen thousand years ago this channel carried water from a melting glacier near Glen Lake across the *Colwood Delta* to the sea. Looking north along Jacklin Road at Belmont Secondary School, you will note that the road descends into another low area, then abruptly rises again to the level of the main delta surface just beyond the B.C. Hydro switchyard. The low area is another large channel that carried meltwater from the vicinity of Langford Lake.

Eastward from the switchyard the form of the meltwater channel, as shown on the map, is that of a tight southerly directed then broad northerly directed pair of meander loops, the latter ending on the Colwood Golf Course near Colwood Corners. The form of the channel is largely obscured by recent construction, but can be clearly recognized on older vertical air photographs. The fact that this channel was not cut deeply into the delta deposits as sea level fell indicates that it was soon abandoned when meltwater from glaciers in the Goldstream watershed began to flow northward into Finlayson Arm, from which glacial ice had retreated.

Continue north along Jacklin Road to Goldstream Avenue (Highway 1A), then turn west (left). This portion of the route is again on the main surface of the delta, where numerous scarps mark the courses of several distributary meltwater channels. Just beyond Spencer School the road dips down toward Langford Lake. Rock outcrops to the north are *Wark Gneiss* and occur on the north side of the *San Juan – Survey Mountain Fault*, which crosses beneath the north margin of the delta to emerge at Fort Rodd Hill Park. The depression occupied by Langford Lake was formed by a large block of stagnant ice, which prevented the deposition of sand and gravel from meltwater streams flowing from the glaciers farther to the west. Nearby, outwash gravels can be seen lying directly on bedrock consisting of *Leech River Complex* strata, occurring between the *San Juan – Survey Mountain* and *Leech River faults.*

Looking westward from near the junction of Goldstream Avenue and the Trans-Canada Highway northwest of Langford Lake,

Viewed to the south from Belmont Secondary School on Jacklin Road in Colwood, the road descends into the valley of a meltwater stream, which, thirteen thousand years ago, carried water from a melting glacier located near Glen Lake across the Colwood Delta to the sea.

you can see a meadow, the level surface of which is interrupted by several small gravel knolls. These knolls, called **kames**, consist of sand and gravel that accumulated in depressions and crevasses on the surface of the melting glacier, and which slumped into place as the ice melted. Some of these kames have been completely excavated as a source of sand and gravel for road building. The meadow is underlain by thick organic soils, which accumulated in the poorly drained hollows between the kames.

From Goldstream Avenue, turn west (left) and travel for a short distance on the Trans-Canada Highway, and then turn west (left) again onto Sooke Lake Road, leading to the Goldstream Provincial Park campsite. This part of the route is through an area of outwash gravels deposited in contact with stagnant glacier ice. Several gravel pits have been excavated in these deposits. At the junction with Humpback Road, the deep gorge to the north was cut by the Goldstream River when meltwater stopped flowing across the Colwood Delta and began to flow northward into Finlayson Arm as sea level descended to its present level. The material that was removed by the river was transported into Finlayson Arm to form the present delta and estuary at the mouth of Goldstream River.

20. THE GULF ISLANDS

The following paragraphs describe the geology of the Gulf Islands as seen from a B.C. ferry passing from Tsawwassen on the mainland to Swartz Bay on Vancouver Island.

The Canadian Gulf Islands, of which there are about fifteen of appreciable size, and many other smaller islands, are formed predominantly from sedimentary strata of the Upper Cretaceous *Nanaimo Group* (see page 29). Exceptions occur on southwestern Saltspring and Portland islands and on most of Moresby Island; these are underlain by Paleozoic rocks of the *Sicker Group* (see page 17). Of the *Nanaimo Group*, conglomerates, sandstones, shales and coal comprise eleven formations, which, in upward stratigraphic order, are the *Comox*, *Haslam*, *Extension*, *Pender*, *Protection*, *Cedar District*, *Decourcy*, *Northumberland*, *Galiano*, *Mayne* and *Gabriola formations*. Throughout the Gulf Islands these strata have been deformed into a series of northwesterly trending folds and faults, which, together with variations in rock type, are responsible for the overall shapes and physiography of the islands and intervening channels.

Upon leaving the terminal at Tsawwassen, the ferry crosses the open Strait of Georgia, a depression carved by southward-moving glaciers and then supplied with sand and gravel outwash deposits as the glaciers melted. Bedrock underlying the glacial deposits forming the floor of the strait, as determined by geophysical techniques, consists of a thick succession of Tertiary nonmarine sediments overlying Cretaceous strata of the *Nanaimo Group*. Beyond the entrance to Active Pass, the cliffs of Galiano Island to the north (starboard side) expose conglomerates and sandstones of the *Gabriola Formation*, which accumulated in a submarine channel that had been cut into underlying strata. As shown in the photograph, notice how the sediments defining the base of the channel seem to dip downwards towards the east and truncate horizontal strata beneath. On the other side of the ship (port side), opposite Mary Anne Point on Galiano Island, which is formed from conglomerate, Miners Bay on Mayne Island is developed upon shales of the *Mayne Formation*. This relationship between promontories composed of resistant sandstone and

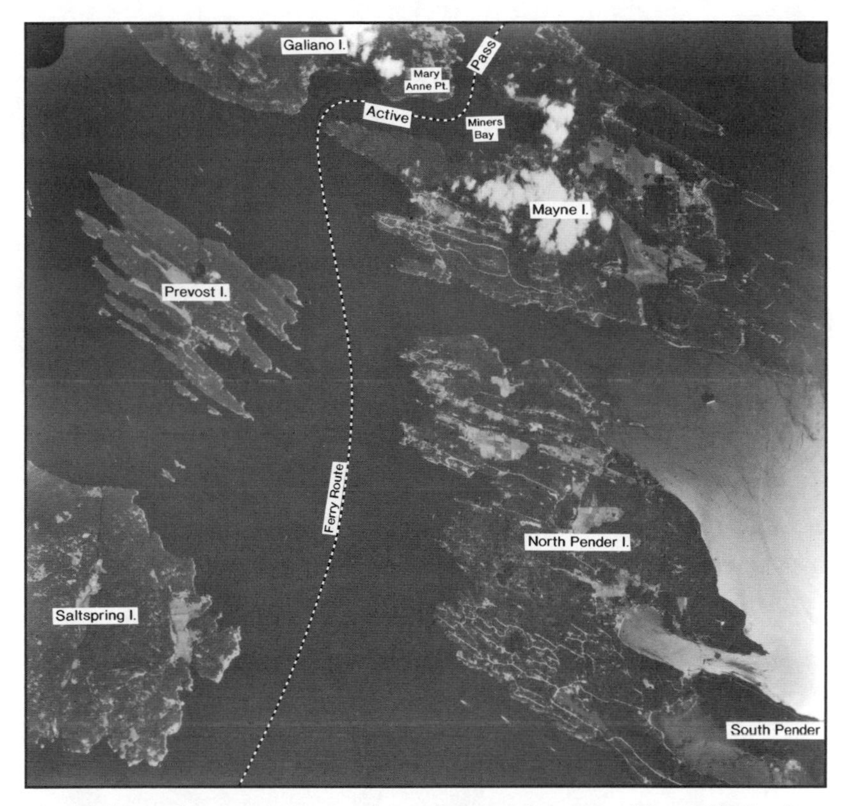

The southeast (lower right) to northwest (upper left) grain of the Gulf Islands is evident in this vertical air photograph taken above Saltspring, Pender, Prevost, Mayne and Galiano islands in the Gulf Islands. The islands are formed from sandstone, conglomerate and shale of the Nanaimo Group, which have been deformed into a series of northwesterly trending folds and faults. The effect of this style of deformation was such as to create a series of narrow promontories and bays, the former formed of resistant sandstone or conglomerate and the latter of more easily eroded shale.

conglomerate, and bays formed within softer shales, is consistent throughout the islands and leads to similarities in physiography and island shape.

After passing between Collinson Point on Galiano Island and Helen Point on Mayne Island, each underlain by conglomerates of the *Galiano Formation*, the ship turns southward to cross Trincomali Channel, beneath which the strata underlying the floor of the channel are folded into a northwesterly trending anticline called the *Trincomali Anticline*. From there the ferry passes between the northwestern end of North Pender Island and the southeastern end of Prevost Island, where the strata have been folded into a syncline. Along the northwestern coast of North

At Mary Anne Point, near the entrance to Active Pass, conglomerate and sandstone of the Upper Cretaceous Gabriola Formation of the Nanaimo Group accumulated in a submarine channel cut into older strata about seventy million years ago.

The waters of Trincomali Channel, between Galiano (right) and Prevost (left) islands, cover strata of the Nanaimo Group that have been folded into a northwesterly trending anticline.

Pender Island, between Port Washington and Mouat Point, several outcrops of the *Galiano* and *Decourcy formations* show characteristic features of submarine landslides.

Continuing southward, the ferry passes by Beaver Point on the southeastern end of Saltspring Island (starboard side). There, sandstone strata of the *Comox Formation*, the lowermost formation of the Cretaceous *Nanaimo Group*, overlie intensely deformed rocks of the Paleozoic *Sicker Group*. The absence of rocks of intervening late Paleozoic and early and middle Mesozoic ages is called an unconformity, meaning that strata of those ages were never deposited here, or if they were deposited, they were subsequently eroded away prior to accumulation of the *Comox Formation* (see Locality 10).

From Beaver Point, the ferry most commonly passes by Portland Island on its port side. This beautiful island, designated Princess Margaret Marine Park, is very popular with the boating community. From the anchorage on its southeastern side, visitors have the choice of walking across well-exposed Cretaceous strata of the *Nanaimo Group* along the east coast, or Paleozoic rocks of the *Sicker Group* on its west coast.

From Portland Island the ferry passes between Knapp (port side) and Piers (starboard side) islands, each formed from strata of the *Extension* and *Pender formations*. The ferry terminal at Swartz Bay is constructed upon sandstone of the *Comox Formation* and shale of the *Haslam Formation*.

GLOSSARY

AGMATITE. A fragmented rock resulting from the injection of light-coloured magmas (eg. quartz diorite) into fractures in dark-coloured rocks (eg. gabbro).

ALEXANDER TERRANE. A piece of crust consisting of Proterozoic, Paleozoic and Triassic sedimentary and volcanic rocks that forms much of southeastern Alaska and southwestern Yukon.

AMYGDULES. Cavities in volcanic rocks, formed by escaping gas as the lava cooled, and filled with minerals such as calcite, quartz or a zeolite.

AMPHIBOLITE. See **hornblende**.

ANGULAR UNCONFORMITY. See **unconformity**.

ANTICLINE. See **fold.**

ANTICLINORIUM. A large upwarped structure, several kilometres in width and length, consisting of many lesser folds.

ARGILLITE. As used in this guide, a hard, dense variety of **shale**, resulting from a low degree of metamorphism, insufficient to result in slate.

BASALT. A form of dark-coloured lava, composed chiefly of calcium **feldspar** and **pyroxene**. Basalt and its coarse-grained equivalent, **gabbro**, are the principal kinds of rocks that form the crust beneath the deep oceans. Commonly, basalt occurs in globular, pillow-like masses called **pillow basalt**.

BIOTITE. One of the mica group of minerals, generally black, dark brown or dark green and consisting of potassium, magnesium, iron and aluminum **silicate**. A common constituent of igneous and metamorphic rocks, it also occurs as a detrital mineral in sedimentary rocks.

BRECCIA. As used in this guide, a consolidated mass of broken fragments of volcanic rock, commonly formed from the breaking up of **pillow basalt** to form **pillow breccia**.

CALCITE. Calcium carbonate ($CaCO_3$), the principal constituent of **limestone**, formed by organic precipitation from sea water. It commonly occurs as a cementing material in sedimentary rocks such as sandstone, where it forms by precipitation of calcium carbonate from ground water.

CANADIAN CORDILLERA. The several systems of mountain ranges and plateaux, extending from the International Boundary to the Beaufort Sea and from the western edge of the Interior Plains to the toe of the continental slope.

CARBONIFEROUS PERIOD. See **Paleozoic Era**.

CENOZOIC ERA. That period in the history of the earth from sixty-six million years ago to the present. The Cenozoic Era is divided into two periods: the **Tertiary Period**, which ended about two million years ago, and the **Quaternary Period**. The Quaternary Period is divided into two epochs: the **Pleistocene Epoch**, which ended about ten thousand years ago, and the **Holocene Epoch**, during which we live today.

CHALCOPYRITE. A metallic mineral, composed of copper, iron and sulphur ($CuFeS_2$), and one of the most important sources of copper.

CHERT. See **radiolarian chert**.

CHILLED MARGIN. The border or marginal area of an igneous intrusion such as a **sheeted dyke**, characterized by smaller crystals than in its interior, and caused by more rapid cooling of the **magma** in the region close to its contact with cold rock.

CHLORITE. A hydrous, greenish mineral, resembling mica, consisting of various combinations of magnesium, iron, aluminum and silica and common in low-grade metamorphic rocks.

CIRQUE. A steep-walled, amphitheatre-like recess occurring at high elevations on the side of a mountain, commonly at the head of a glacial valley, and formed through erosion by a mountain glacier.

CLEAVAGE. The tendency of a rock to split along planes of **foliation**; a property resulting from metamorphism or deformation.

COMPLEX. An assemblage of many different types of rocks whose interrelationships are so structurally complicated as to defy separation into recognizable and mappable formations.

CONCRETION. A hard, compact, spherical or subspherical mass of mineral matter, formed by precipitation from water about a nucleus, such as a shell fragment, and embedded in a porous sedimentary rock, such as sandstone or shale.

CONGLOMERATE. A coarse-grained sedimentary rock, composed of rounded fragments of rock larger than two millimetres in diameter, in a matrix of **sandstone** or **siltstone**. The rock equivalent of gravel.

CONTINENTAL CRUST. See **crust**.

CONTINENTAL DRIFT. The general theory describing the motions of the earth's continents.

CONTINENTAL SHELF. That part of the submerged edge of the continent extending from the coast to the top of the **continental slope** at a depth of about 200 m.

CONTINENTAL SLOPE. The relatively steep slope descending from the edge of the continental shelf to the floor of the deep ocean.

COQUINA. A deposit consisting of concentrated whole and/or fragmented shells.

CRAG-AND-TAIL. Small, narrow elongated ridges, resulting from glaciation of bedrock, having steep, blunt upstream ends (crag) and smooth, tapering downstream surfaces (tail).

CRESCENT TERRANE. A piece of oceanic crust, formed about fifty-four million years ago. On Vancouver Island it consists of lavas and gabbros of the Metchosin Igneous Complex and occurs south of the Leech River Fault. The Crescent Terrane is also known to occur beneath Tertiary sediments underlying the continental shelf off the west coast of the island (see diagram on page 46).

CROSS-BEDDING (Cross-Stratification). Secondary strata inclined at an angle to the main stratification, commonly formed by migrating dunes or ripples.

CRUST. The outermost layer of the earth. Continental crust is approximately 20 to 90 km thick, whereas oceanic crust is about 7 km thick.

DEVONIAN PERIOD. See **Paleozoic Era**.

DIFFERENTIAL EROSION. Erosion occurring at irregular or varying rates, caused by differences in the resistance and hardness of rocks.

DIP. The inclination of strata and other planar geological features such as faults, measured downward from the horizontal.

DIORITE. An **igneous**, **intrusive** rock composed of **plagioclase feldspar**, **hornblende**, **pyroxene** and a small amount of **quartz**.

DRUMLIN. A low, smoothly rounded, elongated hill composed of compact glacial **till** or other glacial deposits, built beneath the edge of moving glacial ice and shaped by the glacial flow. Its long direction is parallel to the direction of flow. Its upstream slope is blunt as compared with the gentler slope of its downstream end.

DYKE. A tabular body of igneous rock that has been intruded into and at an angle to the stratigraphic layering of its host rocks.

EPICENTRE. The point on the earth's surface that is directly above the **focus** of an earthquake.

EPIDOTE. A yellowish-green mineral consisting of calcium, aluminum, iron and silica, formed at low metamorphic temperatures and pressures.

ERRATIC (Glacial). As used in this guide, a large boulder that has been carried to its location by glacial ice and, most commonly, differs in rock type from the bedrock or materials upon which it rests.

FAULT. Faults are fractures in the earth along which rocks on one side of the fracture have moved relative to rocks on the other side. They vary in size and significance from small fractures, called **joints**, where the rocks on either side have moved very little, to enormous structures which separate entire masses of the earth's crust that have moved hundreds to thousands of kilometres.

There are three main types of faults, each type distinguished by the way in which the rocks on either side have moved relative to one another. **Normal faults** are fractures separating blocks of rock where one side has moved downward relative to the block on the other side, and where the fault surface is inclined, or **dips**, toward the downdropped block. In most cases the dip of normal fault surfaces is steep, in the order of 70°, and their **surface traces** (fault lines) vary from a few metres to several tens of kilometres in length. The amount of relative downward displacement also is widely variable and can be up to several kilometres; however, a range from less than a metre to hundreds of metres is most common.

REVERSE FAULTS are fracture surfaces that separate blocks of rocks where one side has moved upwards and over top of rocks on the other side, and where the fault surface is inclined downwards towards the upthrown block. Of the two varieties, the steeper are called **"high angle" reverse faults**, where the amount of displacement and fault lengths are similar to the range embraced by normal faults. The second variety is called a **thrust fault**. These have low inclinations, 20 to 40° or so, and can be many tens or even hundreds of kilometres in length. Moreover, the relative distances the blocks above the fault have moved can range from less than a kilometre to hundreds of kilometres.

The third type of fault is that which separates blocks of rock that have moved horizontally past one another. These are called **wrench** or **strike-slip faults**, the latter term derived from the sense of motion of the opposing blocks, i.e. the blocks move along the trend, or **strike**, of the **surface trace** of the fault. The inclination, or dip, of wrench faults is steep to vertical. Their lengths can range from a few kilometres to many hundreds of kilometres, such as the *Tintina-Northern Rocky Mountain Trench Fault*, which has a surface-trace length (fault-line length) of about 1,200 km and which separates blocks that have moved up to 750 km. A variety of strike-slip fault is a **transform fault**, most of which occur in deep-ocean basins where they offset segments of mid-ocean spreading ridges. One of the most famous is the *San Andreas Fault* of California, which passes through the San Francisco Bay area, and which separates the northward-moving Pacific Plate upon which the people of San Francisco live from the North American Plate and the city of Berkeley. Although not important on Vancouver Island, a prominent transform fault, called the *Queen Charlotte Fault*, very similar to the San Andreas, occurs off the west coast of the Queen Charlotte Islands, where it separates the Pacific and North American plates.

FELDSPARS. A group of closely related minerals composed of silica, oxygen, aluminum and one or more of potassium, sodium and calcium. The feldspars are the most abundant of all minerals, formed by crystallization from molten **magma**. They are the most common constituents of igneous rocks and occur as detrital minerals in sedimentary rocks. The most common types are **plagioclase**, a sodium/calcium aluminum siliate, and **orthoclase**, a potassium/aluminum silicate.

FIORD. A long, narrow, commonly deep, marine coastal inlet, formed by glacial-deepening of a former river valley.

FOCUS. The point of first rupture of an earthquake directly below the **epicentre**.

FOCAL DEPTH. The depth below the **epicentre** (surface) to the point within the earth that is the centre of first rupture (**focus**) of an earthquake.

FOLD. Warped or bent strata. **Anticlines** are up-folds in which originally flat or horizontal strata have been bent to form a convex curve (arch). Conversely, **synclines** are down-folds that form concave curves (basins).

FOLIATION. Laminated structure in a rock resulting from the segregation of different minerals into parallel layers as a result of metamorphism. Commonly, the result of increased pressure and temperature is the flattening of constituent minerals, causing the rock to cleave along fracture-planes parallel to these layers.

FORESET BEDS. The series of inclined layers accumulated as a consequence of sediment being deposited on the steep frontal slope of a delta.

FORMATION. Stratified rocks are grouped according to many criteria, including their composition and location. The most fundamental type of subdivision is the formation, which refers to a succession of strata with specific characteristics and which can be recognized as distinctly different from other formations throughout its region of occurrence. In some instances, several formations sharing common origins or other characteristics are collectively included in a **group** (of formations); examples on Vancouver Island include the Duck Lake, Nitinat and McLaughlin Ridge formations of the Sicker Group.

FRASER GLACIATION. The last principal glacial advance of the **Wisconsinan glacial stage**, which occurred between 29,000 and 10,000 years ago.

GABBRO. A dark-coloured igneous rock in which the proportions of calcium **feldspar**, **pyroxene** and other minerals are within specified limits. Gabbro is the coarse-grained, plutonic equivalent of **basalt**.

GEOLOGICAL ASSOCIATION OF CANADA. A national association of geoscientists, the purpose of which is to foster the practice, teaching and development of Canadian geoscience. It consists of approximately three thousand members, distributed throughout universities, industry and government agencies.

GEOLOGICAL SURVEY OF CANADA (GSC). Canada's largest scientific research organization and one of the oldest and most prestigious of its kind in the world. Since its establishment in 1842, the GSC has provided Canadians with an understanding of the geological architecture of the country, with a view to providing guidance in the responsible development and use of its mineral and energy resources, as well as a knowledge of its natural hazards. With headquarters in Ottawa, it has regional offices in Dartmouth, Nova Scotia, Quebec City, Calgary, Alberta, and in Vancouver and Victoria, British Columbia.

GNEISS. See metamorphic rocks.

GRANODIORITE. An igneous rock in which the proportions of **plagioclase** and **orthoclase feldspar**, **quartz**, **hornblende** and **biotite** are within specified limits.

GROOVES (Glacial). Deep, wide, usually straight furrows cut in bedrock by the abrasive action of rock fragments embedded into the bottom of a moving glacier.

GROUP. See formation.

HOLOCENE. See **Cenozoic Era**.

HORNBLENDE. A common black, dark-green or brown mineral composed of a variable combination of calcium, sodium, magnesium, iron, aluminum, silica and oxygen. It is a common constituent of igneous and metamorphic rocks, the latter including **amphibolite**, a rock consisting almost entirely of hornblende and **plagioclase feldspar**.

IGNEOUS ROCKS. Rocks formed from the solidification of molten **magma**. Those which formed below the earth's surface are coarsely crystalline and are called **intrusive** rocks. When intrusive rocks are uplifted and uncovered by erosion they are called **plutons**. Igneous rocks that solidified at the earth's surface are called extrusive or volcanic rocks; they are finely crystalline.

ILMENITE. Iron, titanium oxide ($FeTiO_3$). A common mineral in **gabbro** and commonly found in sedimentary rocks such as **sandstone**. In some areas the concentration of ilmenite in beach sands is great enough to be mined as a source of titanium.

INCLINATION (Magnetic). The amount of inclination from the horizontal of a vertically oriented dip needle. At the poles the inclination is 90° and at the equator it is 0°. With the use of specialized **magnetometers**, it is possible to determine the inclination of the remanent magnetism in rocks and thus calculate the latitude at which they formed.

INSULAR BELT. The westernmost of five, northwest to southeast oriented subdivisions of the **Canadian Cordillera**. The belts are distinguished from one another according to dominant rock types, physiography and geological history. The Insular Belt includes Vancouver Island, the Queen Charlotte Islands and the St. Elias Mountains of southwestern Yukon and northeastern British Columbia.

INSULAR SUPERTERRANE. A giant terrane consisting of the combined **Alexander Terrane** and **Wrangellia**.

INTRUSIVE ROCKS. See **igneous rocks**.

INTENSITY (Earthquake). A measure of the degree to which an earthquake affects people, property and the ground at a given site within the region wherein an earthquake is felt.

IRONSTONE CONCRETION. A hard, compact, spherical to subspherical mass, commonly up to tens of centimetres in diameter, formed by precipitation of iron oxide about a nucleus (commonly a fossil fragment); commonly found in shales.

ISLAND ARC AND VOLCANIC ARC. Chains of volcanoes constructed above subducting oceanic crust (see **subduction**). As oceanic crust is subducted, it reaches depths where the temperature is sufficient to melt the subducting crust, from where molten material rises upward through the overriding plate to appear at the surface as a chain of volcanoes. In the oceans these chains are called island arcs, and on land they are called volcanic arcs.

ISOSEISMAL MAP. A map showing the regional variation in **intensity** caused by an earthquake.

JOINT. Fractures, commonly occurring in two or more sets, on either side of which there has been little appreciable movement of the rocks.

JURASSIC PERIOD. See **Mesozoic Era**.

KAME. Mounds, knobs, hummocks or knolls composed of stratified glacial sediments that accumulated in depressions on the surface of a melting glacier.

KETTLE. A steep-sided, bowl-shaped depression in glacial drift, commonly without surface drainage, formed by the melting of detached blocks of stagnant ice left behind by a melting and retreating glacier.

LAG DEPOSIT. A residual accumulation of gravel remaining after stream, ocean or air currents have swept away the finer particles.

LIMESTONE. A sedimentary rock formed by the organic precipitation of calcium carbonate ($CaCO_3$).

LODE. A mineral deposit in rock, as opposed to **placer** deposits in stream gravels.

LONGSHORE CURRENTS. A current flowing parallel to the shore, caused by the approach of waves obliquely to the coast.

MAGMA. Molten material, mainly derived from the mantle and intruded into the crust, from which igneous rocks are formed.

MAGNETIC FIELD. The earth's magnetic field is characterized by lines of variable magnetic intensity or force, originating in the earth's core, encircling the globe and focusing near the poles. The field is analogous to that produced by a bar magnet located at the earth's centre and aligned approximately parallel to the rotational axis.

MAGNETITE. Iron, magnesium oxide ($(Fe,Mg)Fe_2O_3$). A common constituent of igneous and sedimentary rocks, and locally an important ore of iron. High concentrations of magnetite occur on several beaches bordering the Strait of Juan de Fuca, where, together with ilmenite, they form so-called "black sands."

MAGNETOMETER. A device for measuring the strength and direction of the earth's magnetic field. Specialized forms of magnetometers are used to measure the remanent magnetism in ancient rocks.

MAGNITUDE (Earthquake). The Richter magnitude scale is based upon the response to ground motion of a "standard seismograph" located 100 km from the epicentre. The scale is logarithmic, meaning that the recorded amplitude of a magnitude seven earthquake is ten times greater than one of magnitude six, one hundred times greater than one of magnitude five, and so on.

MARBLE. Metamorphosed limestone.

MESOZOIC ERA. That interval of earth history occurring between the end of the **Paleozoic Era**, 245 million years ago, and the beginning of the **Cenozoic Era**, 66 million years ago. The Mesozoic Era is divided into three periods: the Triassic Period (245 - 208 million years), the Jurassic Period (208 - 135 million years) and the Cretaceous Period (135 - 66 million years).

METAMORPHIC ROCKS. Rocks that, since their initial formation, have sustained sufficient increase in temperature and pressure such that their original minerals and textures have been changed to new minerals and new textures. **Slate** is the metamorphic equivalent of **shale**, and can be split into thin slabs. **Schist** is a **foliated**, crystalline rock that can be split into slabs due to the parallelism of the minerals present. **Gneiss** is a foliated rock that is commonly banded due to alternating layers of dark- and light-coloured minerals. It does not readily split into slabs.

MIDDEN. A heap or stratum of human refuse (broken pots, tools, ashes, food remains, etc.) found at the former site of an ancient settlement or encampment.

MID-OCEAN RIDGE. A rugged ridge on the deep sea floor, with a central rift valley located along a zone of fractures into which new crustal material is injected from the underlying mantle.

MIGMATITE. Veined, metamorphosed igneous rocks resulting from partial melting of rocks at or near the base of the crust.

MONADNOCK. A conspicuous rocky hill rising above the general level of the surrounding terrain.

MUSCOVITE. A white variety of mica, composed of potassium and aluminum **silicate**; a common constituent of metamorphic rocks.

NORMAL FAULT. See **fault**.

OPHIOLITE. A mass of oceanic crust and underlying mantle that has been thrust onto a continent; the process is opposite of **subduction**.

OLIVINE. An olive-green **silicate** mineral composed of variable proportions of magnesium and iron; a common constituent of **basalt** and **gabbro**. The dark-coloured beach sands of the Hawaiian Islands are rich in olivine.

ORTHOCLASE. See **feldspars**.

PACIFIC RIM TERRANE. A piece of crust consisting of metamorphosed Jurassic and Cretaceous sedimentary and volcanic rocks, separated from **Wrangellia** by the Westcoast and San Juan-Survey Mountain faults and from the **Crescent Terrane** by the Leech River Fault.

PALEOZOIC ERA. That time of earth history between 590 million years ago and the beginning of the Mesozoic Era, 245 million years ago. Divided into six periods: Cambrian (590 - 505 million years), Ordovician (505 - 438 million years), Silurian (438 - 408 million years), Devonian (408 - 360 million years), Carboniferous (360 - 286 million years) and Permian (286 - 245 million years).

PERMIAN PERIOD. See **Paleozoic Era**.

PHYSIOGRAPHIC REGION. A region whose pattern of relief and land forms differs from that of adjacent regions.

PILLOW BASALT. See **basalt**.

PLACER. A deposit of minerals, eroded from bedrock sources and transported and concentrated by streams or waves.

PLAGIOCLASE. See **feldspars**.

PLATE TECTONICS. The corollary hypothesis to **sea-floor spreading**, which explains the motions of the world's crustal plates.

PLEISTOCENE. See **Cenozoic Era**.

PLUTON. As used in this guide, a body of coarsely crystalline granitic rock, originating from solidification of molten magma intruded into the crust and later elevated and exposed by erosion at the surface. Such rocks are called plutonic rocks.

POTHOLE. As used in this guide, a pot-shaped pit or hole eroded into bedrock by the erosive action of stones moved in a swirling fashion by fast-moving water.

PRIMARY WAVE (P Wave). A compressional wave resulting from an earthquake whereby particles move parallel to the direction of motion of the wave.

PYROXENES. A group of closely related dark-coloured silicate minerals composed of one or more of the elements calcium, sodium, magnesium, iron, chromium, manganese or aluminum.

QUARTZ. Next to **feldspar**, the most common rock-forming mineral. Composed of silicon and oxygen (SiO_2) and occurring in igneous, metamorphic and sedimentary rocks such as sandstone.

QUATERNARY PERIOD. See **Cenozoic Era**.

RADIOLARIAN CHERT. Chert is an extremely hard, dense, sedimentary rock consisting of interlocking, microscopic grains of quartz. Commonly it contains the remains of radiolaria, which are microscopic, protozoan planktonic creatures whose skeletons are composed of silica.

RAVELLING. Refers to the manner by which particles of non-cohesive, unconsolidated material such as sand and gravel, forming bluffs and cliffs, flow downslope.

RELIEF. The vertical difference in elevation between hilltops or mountain peaks and the adjacent lowlands or valleys.

REVERSE FAULT. See **fault**.

RICHTER SCALE. See **magnitude**.

ROCHE MOUTONNÉE. An elongated knoll of bedrock sculpted by a moving glacier such that its long direction is oriented parallel to the direction of glacial motion. Its upstream end is generally gently inclined, smoothed and striated or grooved. Its downstream end is steep, rough and blocky.

SANDSTONE. As used in this guide, a sedimentary rock consisting mainly of grains of quartz together with lesser amounts of feldspar and other minerals and rock fragments. The rock equivalent of sand.

SCHIST. See **metamorpic rocks**.

SEA-FLOOR SPREADING. The hypothesis that explains the creation of oceanic crust by convective upwelling of molten magma along the global **mid-ocean ridge** system and the growth, or movement away from the ridge system, of the newly forming crust.

SECONDARY WAVE (S Wave). A shear wave resulting from an earthquake, whereby particles move perpendicularly to the direction of motion of the wave.

SEDIMENTARY ROCKS. Rocks resulting from the consolidation of loose sediment that has accumulated in layers, or strata. The sediments may be produced by mechanical means (fragments of older rocks transported from a source and deposited in water or from air), chemical means (precipitates from solution) or organic means (e.g. limestones constructed from the remains or secretions of plants and animals).

SHALE. A sedimentary rock in which the constituent particles are predominantly of fine silt and clay size; the rock equivalent of mud.

SHEETED DYKES. Tabular sheets formed from the intrusion of **gabbro** into the central rift of a mid-ocean ridge, resulting in a layer of vertically inclined **dykes** spreading outward from the ridge; a consequence of **sea-floor spreading**.

SILICATE. Any mineral whose components include silicon dioxide (SiO_2). Most rock-forming minerals in the earth's **crust** are silicates.

SILTSTONE. A sedimentary rock composed of silt-sized particles (between $1/16$ and $1/256$ mm) of quartz, feldspar and other minerals and rock fragments. The rock equivalent of silt.

SLATE. See **metamorphic rocks**.

SLICKENSIDES. A polished and smoothly striated surface resulting from friction due to faulting.

STRIATIONS (Glacial). Long, thin, finely cut, straight, parallel scratches inscribed into bedrock by the rasping action of rock fragments embedded into the bottom of a moving glacier.

STRIKE. The trend of the intersection of a planar surface (eg. stratum or fault surface) with the horizontal, or the trend or direction of a horizontal line drawn on a planar surface.

STRIKE-SLIP FAULT. See **fault**.

STRUCTURE. The form and architectural characteristics of the earth's features produced by deformation and displacement of rocks (eg. folds and faults).

STRUCTURAL STYLE. The form and trend of structures within a given region (e.g. the structural style of the Gulf Islands is one of northwesterly trending folds and thrust faults).

SUBDUCTION (Subduction Zone). The process of consumption of oceanic crust at the deep sea trenches whereby one piece of crust descends, or is **subducted**, beneath another. An example occurs off our west coast, where the oceanic rocks of the Juan de Fuca Plate, created at the Juan de Fuca Ridge, have spread away from the ridge through **sea-floor spreading** and are being consumed along the Cascadia **subduction zone** off the west coast of Vancouver Island, Washington, Oregon and northernmost California.

SUBMARINE TRENCH. Linear, deep-sea, narrow valleys formed at the junction of converging tectonic plates and caused by subduction of one plate beneath the other.

SURFACE TRACE of a fault (Fault Line). The line of intersection of a fault surface with the ground surface.

SYNCLINE. See **fold**.

TERRANES. Parts of the earth's crust which preserve geological records different from those of neighbouring terranes. The boundaries between terranes are major faults. Some terranes appear to be comparatively thin sheets, whereas others are at least 18 to 20 km thick. Some terranes originated close to their present position whereas others were formed thousands of kilometres from their current locations.

TERTIARY PERIOD. See **Cenozoic Era**.

THRUST FAULT. See **fault**.

TILL. Unsorted and unstratified glacial drift deposited by and beneath a glacier; generally consisting of clay, silt, sand, gravel and boulders.

TOPSET BEDS. Nearly horizontal layers of sediment deposited on the top surface of an advancing delta; topset beds commonly overlie **foreset beds**.

TRANSFORM FAULT. See **fault**.

TRIASSIC PERIOD. See **Mesozoic Era**.

TUFF. As used in this guide, a rock composed of consolidated volcanic ash or other very fine-grained volcanic debris.

U-SHAPED VALLEY. A valley with a pronounced U-shape in cross-section, with steep walls and a broad, nearly flat floor, carved by an advancing tongue of glacial ice.

UNCONFORMITY. A time-gap in the geological record represented by an absence of rock due to a period of erosion or nondeposition. For example, on Vancouver Island, where rocks of the Lower Jurassic Bonanza Group (approximately 183 million years old) are overlain by strata of the Upper Cretaceous Nanaimo Group (approximately 85 million years old), the contact between them is an unconformity representing a time hiatus of approximately 100 million years duration. An **angular unconformity** is one separating two groups of rocks whose strata are not parallel; usually younger strata above the unconformity rest upon the eroded surface of tilted or folded older strata beneath.

VESICLES (VESICULAR). Pores in a volcanic rock caused by the expulsion of gas during cooling of molten lava (see **amygdules**).

VOLCANIC SANDSTONE. A sedimentary deposit of sand-sized volcanic grains eroded from other volcanic rocks such as lava.

WAVE-CUT TERRACE. A gently sloping surface, produced by wave erosion, extending seaward from the base of a wave-cut cliff.

WISCONSINAN GLACIAL STAGE. See **Fraser Glaciation**.

WRANGELLIA. That fragment of earth's crust consisting of the Paleozoic, Triassic and Jurassic rocks of Vancouver Island, the Queen Charlotte Islands and parts of southwestern Yukon, northwestern British Columbia and southeastern Alaska.

WRENCH FAULT. See **fault**.

ZEOLITES. A general term for a large group of hydrous aluminum silicates, similar in composition to feldspars, with sodium, calcium and potassium as their chief metals.

SOURCES OF ADDITIONAL INFORMATION

Geological Survey Branch
B.C Ministry of Energy, Mines
and Petroleum Resources
5th Floor, 1810 Blanshard Street
Victoria, B.C. V8V 1X4
604/952-0374
Library on 1st floor
Publication Sales:
Crown Publications Inc.
521 Fort Street
Victoria, B.C. V8W 1K6
604/386-4636

Cordilleran Division
Geological Survey of Canada
100 West Pender Street
Vancouver, B.C. V6B 1R8
604/666-0529
Library on 5th floor
Publications Sales:
6th Floor; 604/666-0271

Department of Geophysics and Astronomy
University of British Columbia
2075 Westbrook Place
Vancouver, B.C. T2N 1N4
604/822-2267

Pacific Geoscience Centre
Geological Survey of Canada
P.O. Box 6000
Sidney, B.C. V8L 4B2
604/363-6500

School of Earth and Ocean Sciences
E-Hut; P.O. Box 1700
University of Victoria
Victoria, B.C. V8W 2Y2
604/721-8848

Department of Geological Sciences
University of British Columbia
6339 Stores Road
Vancouver, B.C. V6T 2B4
604/822-2449

ADDITIONAL READING

The following list of scientific papers and publications on Vancouver Island geology is far from complete. Those in bold type are considered by the authors to be the most useful as additional information pertaining to the geology of southern Vancouver Island. Publications of the Geological Survey Branch of the British Columbia Ministry of Energy, Mines and Petroleum Resources and the Geological Survey of Canada are available from the sources listed under "Additional Sources of Information"; some of these are out of print but are available at university libraries.

Magazines, such as *National Geographic*, *Scientific American* and others, contain articles written in a more popular form than those in the research publications listed below. The interested reader can access articles on such topics as plate tectonics, glaciation, volcanism, etc., through computerized indices of periodical literature in the public libraries of larger cities, where most of the important magazines are held in their collections.

GLACIAL GEOLOGY

Alley, N.F. and Chatwin, S.C., 1979. Late Pleistocene history and geomorphology, southeastern Vancouver Island, B.C. *Canadian Journal of Earth Sciences*, 16: 1645-1657.

Clague, J.J., 1976. Quadra Sand and its relation to late Wisconsin glaciation of southwest British Columbia. *Canadian Journal of Earth Sciences*, 13: 803-815.

Clague, J.J., Harper, J.R., Hebda, R.J. and Howse, D.E., 1982. Late Quaternary sea levels and crustal movements, coastal British Columbia. *Canadian Journal of Earth Sciences*, 19: 597-618.

Clague J.J., 1989. "Quaternary geology of the Canadian Cordillera." Chapter 1 in

***Quaternary Geology of Canada*, R.J. Fulton (ed.). Geological Survey of Canada, Geology of Canada, no. 1 (also *Geological Society of America, The Geology of North America, K-1*), 17-96.**

MINERAL DEPOSITS

Fyles, J.T., 1949. Copper deposits of the Sooke Peninsula, British Columbia. British Columbia Ministry of Mines, Annual Report for 1948, A162-A170.

Hora, Z.D. and Miller, L.B., 1994. Dimension stone in Victoria, B.C.; British Columbia Ministry of Energy, Mines and Petroleum Resources, Geological Survey Branch, Information Circular No. 1994-15.

Neumann, N., 1991. The geology and history of mining at Goldstream; a report prepared for the Pacific Section of the Geological Association of Canada. Available at the Greater Victoria Public Library; the library of the Geological Survey Branch of the B.C. Ministry of Energy, Mines and Petroleum Resources, 1810 Blanshard Street; and the McPherson Library at the University of Victoria.

PALEONTOLOGY

Bell, W.A., 1957. Flora of the Upper Cretaceous Nanaimo Group of Vancouver Island, British Columbia. Geological Survey of Canada, Memoir 293.

Cameron, B.E.B., 1980. Biostratigraphy and depositional environment of the Escalante and Hesquiat formations (Early Tertiary) of the Nootka Sound area, Vancouver Island, British Columbia. Geological Survey of Canada, Paper 78-9.

Ludvigsen, R. and Beard, G., 1994. Westcoast fossils: A guide to the ancient life of Vancouver Island. Whitecap Books, Vancouver and Toronto.

McGugan, A., 1962. Upper Cretaceous foraminiferal zones, Vancouver Island, British Columbia. *Journal of the Alberta Society of Petroleum Geologists*, 10: 585-592.

McGugan, A., 1964. Upper Cretaceous zone foraminifera, Vancouver Island, British Columbia, Canada. *Journal of Paleontology*, 38: 933-951.

McGugan, A., 1979. Biostratigraphy and paleoecology of Upper Cretaceous (Campanian and Maestrichtian) foraminifera from the Upper Lambert, Northumberland and Spray Formations, Gulf Islands, British Columbia, Canada. *Canadian Journal of Earth Sciences*, 16: 2,263-2,274.

McGugan, A., 1981. Late Cretaceous (Campanian) foraminiferal faunas, Charter et al. Saturna No. 1, Gulf Islands, British Columbia; *Bulletin of Canadian Petroleum Geology*, 29: 110-117.

Usher, J.L., 1952. Ammonite faunas of the Upper Cretaceous rocks of Vancouver Island. Geological Survey of Canada, Bulletin 21.

Yole, R.W., 1963. An Early Permian fauna from Vancouver Island, British Columbia. *Bulletin of Canadian Petroleum Geology*, 11: 138-149.

POPULAR GEOLOGY

Weston, J., 1986. "Landscapes of time around Victoria" in *The Naturalist's Guide to the Victoria Region*; Victoria Natural History Society, 15-70.

Yorath, C.J., 1990. *Where Terranes Collide*. Orca Book Publishers, Victoria, B.C.

REGIONAL GEOLOGY

Carson, D.J.T., 1973. The plutonic rocks of Vancouver Island; Geological Survey of Canada, Paper 72-44.

Clapp, C.H., 1912. Southern Vancouver Island. Geological Survey of Canada, Memoir 13.

Clapp, C.H. and Cook, H.C., 1917. Sooke and Duncan map-area, Vancouver Island, British Columbia. Geological Survey of Canada, Memoir 96.

Fyles, J.T., 1955. Geology of the Cowichan Lake area. British Columbia Department of Mines, Bulletin 37.

Gabrielse, H. and Yorath, C.J., eds., 1992. *The Cordilleran Orogen in Canada*. Geo-

logical Survey of Canada, Geology of Canada, no. 4 (also Geological Society of America, The Geology of North America, v. G-2). 844 p.

Massey, N.W.D., 1986. The Metchosin Igneous Complex, southern Vancouver Island: ophiolite stratigraphy developed in an emergent island setting. *Geology*, 14: 602-605.

Massey, N.W.D., 1993. Geology and mineral resources of the Alberni – Nanaimo Lakes sheet, Vancouver Island 92F/1W, 92F/2E and part of 92F/7E. British Columbia Ministry of Energy, Mines and Petroleum Resources, Geological Survey Branch, Paper 1992-2.

Massey, N.W.D., 1994. Geological compilation, Vancouver Island, British Columbia (NTS 92B,C,E,F,G,K,L, 102I). B.C. Ministry of Energy, Mines and Petroleum Resources, Open File 1994-6, 5 digital files, legend, 1:250,000 scale.

Massey, N.W.D., Friday, S.J., Tercier, P.E. and Potter, T.E., 1991. Geology of the Duncan area (92B/13). British Columbia Ministry of Energy, Mines and Petroleum Resources, Geological Survey Branch, Geoscience Map 1991-3.

Massey, N.W.D., Friday, S.J., Tercier, P.E., Rublee, V.J. and Potter, T.E., 1991. Geology of the Cowichan Lake area, NTS 92C/16. British Columbia Ministry of Energy, Mines and Petroleum Resources, Geological Survey Branch, Geoscience Map 1991-2.

Massey, N.W.D., Friday, S.J., Riddell, J.M. and Dumais, S.E., 1991. Geology of the Port Alberni – Nanaimo Lakes area; British Columbia Ministry of Energy, Mines and Petroleum Resources, Geological Survey Branch, Geoscience Map 1991-1.

Muller, J.A., 1983. Geology of Victoria. Geological Survey of Canada, Map 1553A.

Muller, J.E., Northcote, K.E. and Carlisle, D., 1974. Geology and mineral deposits of Alert – Cape Scott map-area, Vancouver Island, British Columbia. Geological Survey of Canada, Paper 74-8.

Muller, J.E., Cameron, B.E.B. and Northcote, K.E., 1981. Geology and mineral deposits of Nootka Sound map-area, Vancouver Island, British Columbia. Geological Survey of Canada, Paper 80-16.

Rusmore, M.E. and Cowan, D.S., 1985. Jurassic-Cretaceous rock units along the southern edge of the Wrangellia terrane on Vancouver Island. *Canadian Journal of Earth Sciences*, 22: 1,223-1,232.

Stevenson, J.S., 1945. Geology and ore deposits of the China Creek area, Vancouver Island, British Columbia. Ministry of Mines, British Columbia, Annual Report, p. A1437-A161.

Yorath, C.J., comp., In Press: LITHOPROBE, Southern Vancouver Island: Geology. Geological Survey of Canada, Bulletin.

STRATIGRAPHY

Jeletzky, J.A., 1950. Stratigraphy of the west coast of Vancouver Island between Kyuquot Sound and Esperenza Inlet, British Columbia. Geological Survey of Canada, Paper 50-57.

Jeletzky, J.A., 1976. Mesozoic and Tertiary rocks of Quatsino Sound, Vancouver Island, British Columbia. Geological Survey of Canada, Bulletin 242.

Muller, J.E., 1980. The Paleozoic Sicker Group of Vancouver Island, British Columbia, Geological Survey of Canada, Paper 79-30.

Muller, J.E. and Jeletzky, J.A., 1970. Geology of the Upper Cretaceous Nanaimo Group, Vancouver Island and Gulf Islands, British Columbia. Geological Survey of Canada, Paper 69-25.

Pacht, J.A., 1984. Petrologic evolution and paleogeography of the Late Cretaceous Nanaimo Basin, Washington and British Columbia: Implications for Cretaceous tectonics. *Geological Society of America Bulletin*, 95: 766-778.

Ward, P.D., 1978. Revisions to the stratigraphy and biochronology of the Upper Cretaceous Nanaimo Group, British Columbia and Washington State. *Canadian Journal of Earth Sciences*, 15: 405-423.

Yole, R.W., 1968. Upper Paleozoic stratigraphy of Vancouver Island, British Columbia. *Proceedings of the Geological Association of Canada*, 20: 30-40.

STRUCTURAL GEOLOGY

England, T.J.E. and Calon, T.J., 1991. The Cowichan fold and thrust system, Vancouver Island, southwestern British Columbia. *Geological Society of America Bulletin*, 103: 336-362.

Fairchild, L.H. and Cowan, D.S., 1982. Structure, petrology and tectonic history of the Leech River complex northwest of Victoria, Vancouver Island. *Canadian Journal of Earth Sciences*, 19: 1817-1835.

Yorath, C.J., Green, A.G., Clowes, R.M., Sutherland Brown, A., Brandon, M.T., Kanasewich, E.R., Hyndman, R.D. and Spencer, C., 1985a. LITHOPROBE, southern Vancouver Island: Seismic reflection sees through Wrangellia to the Juan de Fuca plate. *Geology*, 13: 759-762.

TECTONICS

Brandon, M.T., 1985. Mesozoic melange of the Pacific Rim Complex, western Vancouver Island; in Field Guides to Geology and Mineral Deposits in the southern Canadian Cordillera, Geological Society of America, Cordilleran Section Meeting, Vancouver, B.C. 1985, Field Trip No. 7.

Clowes, R.M., Brandon, M.T., Green, A.G., Yorath, C.J., Sutherland Brown, A., Kanasewich, E.R. and Spencer, C., 1987a. LITHOPROBE – southern Vancouver Island: Cenozoic subduction complex imaged by deep seismic reflections. *Canadian Journal of Earth Sciences*, 24: 31-51.

Hyndman, R.D., Yorath, C.J., Clowes, R.M. and Davis, E.E., 1990. The northern Cascadia subduction zone at Vancouver Island: Seismic structure and tectonic history. *Canadian Journal of Earth Sciences*, 27: 313-329.

Muller, J.E., 1977. Evolution of the Pacific margin, Vancouver Island and adjacent regions. *Canadian Journal of Earth Sciences*, 14: 2062-2085.

Sutherland Brown, A. and Yorath, C.J., 1985. LITHOPROBE profile across southern Vancouver Island: Geology and tectonics; in *Field Guides to Geology and Mineral Deposits in the Southern Canadian Cordillera*, Geological Society of America, Cordilleran Section Meeting, Vancouver, B.C., 1985, Field Trip No. 8.

INDEX

A (p) following a page number is used to denote a photograph reference either on that page or, when there is a grouping of pages, within those page numbers.

H

I

J

K

L

M

N

O

P

Q

R

S